EAI/Springer Innovations in Communication and Computing

Series Editor

Imrich Chlamtac, European Alliance for Innovation, Ghent, Belgium

The impact of information technologies is creating a new world yet not fully understood. The extent and speed of economic, life style and social changes already perceived in everyday life is hard to estimate without understanding the technological driving forces behind it. This series presents contributed volumes featuring the latest research and development in the various information engineering technologies that play a key role in this process. The range of topics, focusing primarily on communications and computing engineering include, but are not limited to, wireless networks; mobile communication; design and learning; gaming; interaction; e-health and pervasive healthcare; energy management; smart grids; internet of things; cognitive radio networks; computation; cloud computing; ubiquitous connectivity, and in mode general smart living, smart cities, Internet of Things and more. The series publishes a combination of expanded papers selected from hosted and sponsored European Alliance for Innovation (EAI) conferences that present cutting edge, global research as well as provide new perspectives on traditional related engineering fields. This content, complemented with open calls for contribution of book titles and individual chapters, together maintain Springer's and EAI's high standards of academic excellence. The audience for the books consists of researchers, industry professionals, advanced level students as well as practitioners in related fields of activity include information and communication specialists, security experts, economists, urban planners, doctors, and in general representatives in all those walks of life affected ad contributing to the information revolution.

Indexing: This series is indexed in Scopus, Ei Compendex, and zbMATH.

About EAI - EAI is a grassroots member organization initiated through cooperation between businesses, public, private and government organizations to address the global challenges of Europe's future competitiveness and link the European Research community with its counterparts around the globe. EAI reaches out to hundreds of thousands of individual subscribers on all continents and collaborates with an institutional member base including Fortune 500 companies, government organizations, and educational institutions, provide a free research and innovation platform. Through its open free membership model EAI promotes a new research and innovation culture based on collaboration, connectivity and recognition of excellence by community.

Lucia Knapčíková • Dragan Peraković
Editors

7th EAI International Conference on Management of Manufacturing Systems

Springer

EAI
RESEARCH MEETS INNOVATION

Editors
Lucia Knapčíková
Department of Industrial Engineering and Informatics, Faculty of Manufacturing Technologies
Technical Univ Kosice
Presov, Slovakia

Dragan Peraković
Faculty of Transport & Traffic Sciences
University of Zagreb
Zagreb, Croatia

ISSN 2522-8595 ISSN 2522-8609 (electronic)
EAI/Springer Innovations in Communication and Computing
ISBN 978-3-031-22718-9 ISBN 978-3-031-22719-6 (eBook)
https://doi.org/10.1007/978-3-031-22719-6

This Springer imprint is published by the registered company Springer Nature Switzerland AG
The registered company address is: Gewerbestrasse 11, 6330 Cham, Switzerland

Preface

7th EAI International Conference on Management of Manufacturing Systems (EAI MMS 2022), October 5–7, 2022, Krynica-Zdrój, Poland

The EAI MMS 2022 – 7th EAI International Conference on Management of Manufacturing Systems aims to bring together world-leading academics and practitioners from the fields of management, economics, infrastructure planning, and manufacturing. The unique combination of fields and disciplines focuses on Industry 4.0, the power of smarter information, and opportunities to create a bridge between science and practice. We are delighted to introduce the seventh edition of the 2022 European Alliance for Innovation (EAI) International Conference on Management of Manufacturing Systems. This conference has brought together researchers, developers, and practitioners from around the world. In light of the latest knowledge and findings from scientific projects, the authors present actual R&D trends in the given field. This issue defines the state of the art in the area and explores related topics for future research.

The proceeding of EAI MMS 2022 consisted of 18 full papers, including 2 invited papers in oral presentation sessions at the main conference tracks. Aside from the high-quality technical paper presentations, the technical program also featured two keynote speeches, one invited talk, and two technical workshops. The two keynote speakers were Justyna Trojanowska, Ph.D., Eng., from Poznan University of Technology, Poland, with her keynote speech "Using PowerBI to Support Decision-Making in the Production Area" and Dipl.-Ing. Dr.techn. Markus Brillinger from Pro2Future, GmbH., Graz, Austria, with his keynote speech "Sustainable Production Systems – Game Changers from Augmented Reality to Zero-Defect Manufacturing." It was also a great pleasure to work with such an excellent organizing committee team. We are also grateful to Conference EAI Manager Radka Vasileiadis, for her support, and all the authors who submitted their papers to the EAI MMS 2022 conference, and reviewers for their hard work.

Our ambition was to establish communication channels and disseminate knowledge among professionals working in manufacturing and related institutions. We strongly believe that future cooperation will be as successful and stimulating as indicated by the contributions presented in this volume.

Presov, Slovakia Lucia Knapčíková
Zagreb, Croatia Dragan Peraković

Contents

Part I
Smart Techniques/Smart Technology

Concept, Architecture, and Performance Testing of a Smart Home Environment for the Visually Impaired Persons

Marko Periša, Ivan Cvitić, Petra Zorić, and Ivan Grgurević

1 Introduction

Independent moving and living are prerequisites for enhancement in the degree of mobility and quality of life for the visually impaired persons. According to the latest official data, there are currently 19,132 visually impaired persons in the Republic of Croatia [1]. With the development of advanced information and communication technologies and services, it is possible to develop a model of assistive technology to raise the degree of independence in everyday activities. Everyday activities at home where the visually impaired person (user) lives means all the activities such as preparing meals, cleaning, use of different reminders (medication, obligations, etc.), free time (listening to music or watching TV), monitoring of user's health activities (mHealth), light level management, air quality management, air conditioning, and TV and radio device management.

Smart homes have the ability to integrate these user activities and to provide a unique service of managing individual devices and informing the users. The management of devices today is possible through virtual assistants such as Google Assistant, Amazon Alexa, Microsoft Cortana, or Apple Siri that have been already integrated into smartphones. In this way the user is enabled to live more comfortably, safely, healthily, and independently. However, the devices and the *Internet of Things* (IoT) networks that are used in the concept of smart homes are becoming increasingly complex, which can lead to problems while using the service. This alone can result in the disruption of the entire smart home ecosystem. It is, therefore,

M. Periša (✉) · I. Cvitić · P. Zorić · I. Grgurević
University of Zagreb, Faculty of Transport and Traffic Sciences, Zagreb, Croatia
e-mail: marko.perisa@fpz.unizg.hr; ivan.cvitic@fpz.unizg.hr; petra.zoric@fpz.unizg.hr; ivan.grgurevic@fpz.unizg.hr

L. Knapčíková, D. Peraković (eds.), *7th EAI International Conference on Management of Manufacturing Systems*, EAI/Springer Innovations in Communication and Computing, https://doi.org/10.1007/978-3-031-22719-6_1

important to define the taxonomy of user-dependent devices and technologies while creating a smart home for the visually impaired persons.

2 Previous Research

A smart home is "a residence equipped with a communications network, linking sensors, domestic appliances, and devices, that can be remotely monitored, accessed or controlled, and which provides services that respond to the needs of its inhabitants" [2]. According to the data on the quantity of IoT devices used in smart environments until the year 2025, there are 71 billion installed IoT devices expected, whereas in Smart Phone environments 12.86 billion devices are expected [3]. Virtual assistants used to manage IoT devices in smart homes also indicate an increasing role in everyday life activities of the users. By analyzing the characteristics of work and the possibilities of virtual assistants the interaction between people and devices can be enhanced by using different technologies, such as gesture recognition, image/video recognition, speech recognition, and all this with the aim of creating user knowledge base [4].

To visually impaired persons, voice control of IoT devices at homes is imposed as an option where by using one application solution the user has a simpler option of controlling all IoT devices. The IoT agent is an application solution that can understand textual or voice commands using Natural Language Processing (NLP) [5]. While designing mobile applications whose basic role is voice or text management of IoT devices, it is important to use the standards for the visually impaired persons: *Synchronization Accessibility User Requirements* (SAUR) and *Web Accessibility Initiative* (WAI) [6]. Some research emphasize the importance of inclusion in the design of the applications for the people with disabilities and its interaction [7]. User interfaces also require a certain approach in design; therefore, it is important to follow the universal design guidelines [8]. Management of IoT devices in homes from the aspect of energy saving can also take place through customized application solutions for visually impaired persons, as well as for the senior group of users [9]. The smart home system adapted for the visually impaired persons, which is voice-controlled, apart from the device management functionality, also has the ability of providing information from everyday events (news, hours, and weather) [10].

Adapting the smart home environment to the social context and the design of the home environment is also important in the future development of IoT services, all with the aim of performing everyday activities [11]. One of the possibilities of implementing IoT technologies to users is the use of BAN (*Body Area Network*), which with the appropriate service of monitoring user health has the ability of integrating sensors on the clothing [12].

In the field of development of new innovative digital services and the digital transformation in general, ETSI is emerging as a key stakeholder in the development of high-quality ICT standards. ETSI Technology Radar was developed with the aim of creating future standards of the ETSI organization for all challenges that come from the world of industry, transport, smart cities, etc. and all the benefits of modern information and communication technologies and networks [13]. The application of autonomous and intelligent networks and systems is today increasingly used also indoors, mostly in the field of eHealth and in providing services for senior persons and persons with disabilities [14, 15]. The Ambient Assisted Living (AAL) concept whose aim is to assist and provide real-time information to senior persons and persons with disabilities is also increasingly used in designing new services based on the decision support systems [16, 17].

The mentioned research provides an overview of the current possibilities of managing IoT devices, development of advanced information networks and technologies, and the future standards in the field of digital transformation.

3 Smart Home Concept for the Visually Impaired Persons

In order to design the conceptual system architecture, the smart home environment for the visually impaired can be considered through three segments: user needs, home environment or architecture, as well as the technology used in the home [18].

It is important to design a smart home according to the user needs, according to the type of user's impairment (blindness or low vision), user preferences, routine of daily living, and physical abilities. According to a research carried out with the aim of identifying the user needs, as many as 52% of users have the desire to manage IoT devices within their homes by using mobile or web applications [19].

The research included 42 users. According to the data of this research, 56% of users already use one of the applications for managing IoT devices within their homes, which mostly refers to the management of light intensity in the kitchen. The complexity of using several applications on mobile devices leads to physical and mental strain on the user, which additionally burdens the user and leads to a reduced quality of life. Reading of RFID or NFC labels inside the space is also an additional physical effort for blind people, thus excluding the support of the system designed according to the universal design guidelines.

User desires related to the sensor management and provision of information in a smart home, according to the mentioned research, are presented in Fig. 1.

In addition to the presentation in Fig. 1, inside the smart home, the user can be informed by voice about the time (hour and minutes), reminder control, fire detection, vibration (earthquake), flood, etc. Regarding devices, one can control TV and radio devices, refrigerator, vacuum cleaner, air conditioning, front door, shutters, cameras, and other available IoT devices. It is also possible to provide the user with health information with the possibility of monitoring the user's pulse, blood pressure, temperature, amount of sugar, and oxygen in the blood, presented in more detail in Fig. 2.

Figure 2 shows the concept of a smart home adapted for the visually impaired persons (blind and low-vision). Within home there are sensors for safer and more

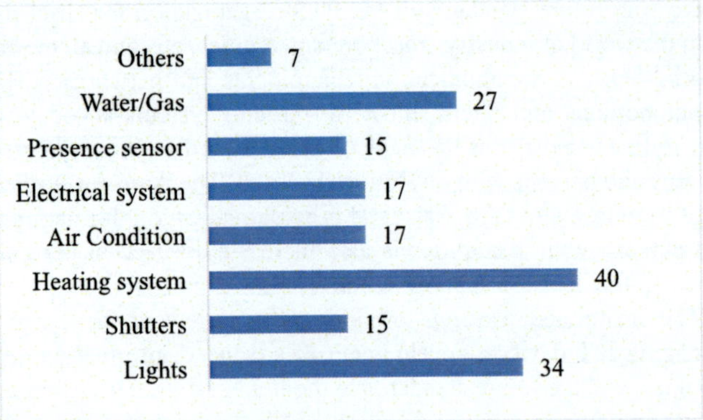

Fig. 1 Automated systems: user preferences [19]

comfortable performance of the daily activities. The virtual assistant offers voice control of smart IoT devices, and provision of information to the user, collected from the sensors. The information from the sensor is delivered at the request of the user. For the mentioned concept it is important to use the application solution made according to the universal design guidelines. The idea of the presented concept is to enable the creation of a user base of the user needs with the aim of providing accurate and real-time information.

4 Conceptual Model of Smart Home System Architecture

For the purpose of delivering the service to the user of the smart home, the communication technology and the elements of the conceptual system architecture have been defined. Also, the basic characteristics of autonomous networks have been presented, that have been recognized as a new possibility of improving the safety of service delivery of accurate and real-time information within the environment of one's own home.

4.1 Taxonomy of Technology and Devices

For the purpose of delivering the service to the user, in this chapter the environment of a smart home has been divided into four layers of automation: Perception and Actuations, Network/Transmission, Preprocessing and Middleware, and Applications as presented in Fig. 3.

Perception and Actuations layer presents physical objects that are used to collect information of interest from the smart home environment, and to transform them

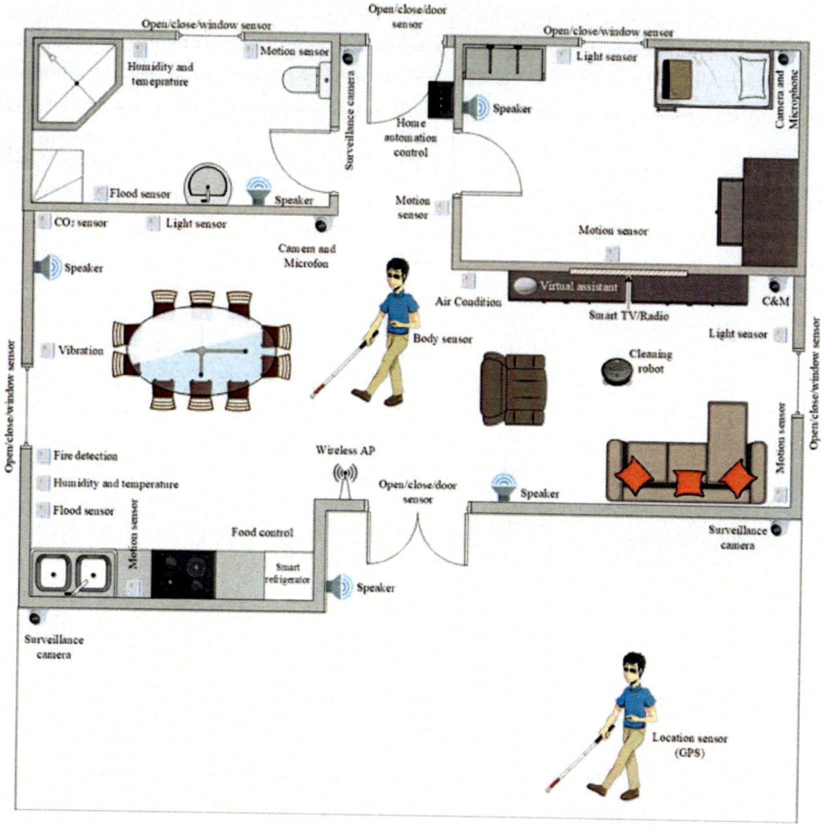

Fig. 2 Concept of smart home for the visually impaired persons

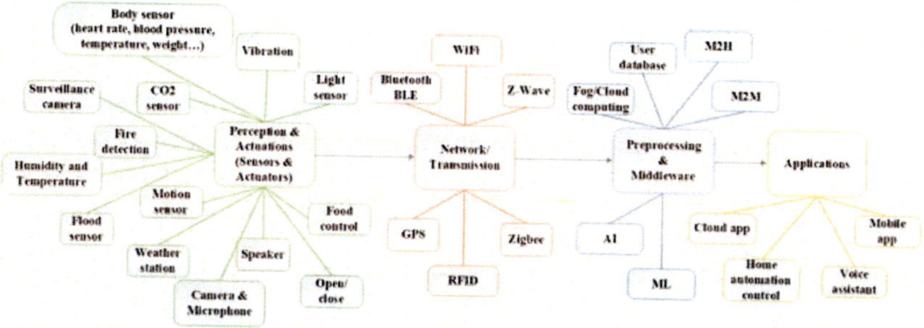

Fig. 3 Taxonomy of the used technology

into digital data. This layer consists of various types of sensors and actuators. A part of the responsibility of this layer is to identify and collect data on the characteristics of each sensor and actuator. The sensors used in this chapter are the sensors that collect data on the user health (heart rate, blood pressure, temperature, weight, etc.), the sensors warning of possible accidents (CO_2 sensor, fire detection, flood sensor), light sensor to monitor the amount of light in a smart home, humidity and temperature sensors that are informative, vibration and motion sensors, surveillance camera, camera and microphone, speaker, weather station, and food control sensor. Open/close sensor (possible for doors and windows) is an actuator used in the solution proposal of this chapter.

Network layer or transmission layer is responsible for secure transmission of data obtained from the sensor devices to the processing system, i.e., to the Preprocessing and Middleware layer through various network technologies. It consists of an access sublayer that has the role of collecting data from the Perception layer and of sending it to the Internet sublayer. In the access sublayer, that is in the part of the core network, the main role is played by the technologies Z-Wave, Zigbee, Bluetooth BLE, GPS, and RFID. WiFi technology is used to transfer the obtained information from the access sublayer to the next layer, Preprocessing and Middleware.

Preprocessing and Middleware layer is a software layer that can directly synchronize the services with the appropriate requirements and has a connection to the database. It performs the processes such as preparing the information for further processing, data cleaning, standardization, values of data within a dataset, balancing, intelligent routing, and translating network addresses. It is based on cloud computing because of the great benefits it provides, such as scalability and flexibility. Fog computing has the role of cloud relief. In the proposal of this chapter, the user data are saved in the user database that is stored in the cloud infrastructure. The preparation of information for delivery in the appropriate form is enabled by the use of M2H (Machine–to–Human) and M2M (Machine–to–Machine) technologies. The ML (machine learning) algorithms enable intelligent adaptation of applications to changes in the environment, and the AI (Artificial Intelligence) technology helps to collect data from the devices, predicts the user behavior within the smart home environment, and helps improve data security.

Based on the processed information in the Preprocessing and Middleware layer, the Application layer is responsible for providing services to end users and the proper operation of applications. The functions that are performed via the Cloud app are accessed using mobile application that is implemented on the user device and the corresponding API (Application Programming Interface). Through home automation control the user can monitor and control certain functions within the smart home environment, such as lighting, air conditioning, various household appliances, and the like.

4.2 *Autonomous Networks*

Autonomous network represents the improvement of a large number of possibilities created by the software-defined networks. According to [20], autonomous networks consist of network and software platforms that, with a minimum of human contribution or even no human contribution at all, can sense the environment in which they are located and adapt their behavior accordingly. Autonomous network goes beyond the basic possibilities of automation and in this way it becomes a network that works on its own. This type of network changes the role that today's networks play in motivating the innovations. Some of the characteristics of autonomous networks are open eco-systems, software-driven infrastructure (including wireless and wired networks, IoT edges, and data centers), and insight and analytics to enable innovations (use of data to analyze the historical traffic, predicting changes of things over time in order to optimize the execution and predict network failures) [21].

Autonomous network adapts dynamically to changes that occur in the network environment, and its infrastructure is based on a self-developing telecommunication network. The goal of autonomous networks is to provide simple and easy solutions of use with "Zero-X" (zero wait services, zero touch maintenance, zero trouble experience) automation based on fully automated operations of the life cycle "Self-X" (self-serving, self-fulfilling, self-assuring) which serves to dynamically adapt to user needs and available resources [22].

The classification system for understanding the network autonomy for autonomous networks builds a maturity model for estimating the level of automation achieved by an autonomous network. There are six levels of network automation from "no automation" to "full automation" [23].

Autonomous networks are already being applied multiple times in the real world. One of the examples is the detection of intrusion into the network by means of ML. Such network functionality is necessary in the smart home environment where there are multiple devices that are interconnected and where security breaches can easily occur. Autonomous networks in this scenario search for anomalies such as credential stuffing attacks and they block the attempt to attack the endpoints that are invisible to traditional security tools.

4.3 Conceptual System Architecture

In order to deliver all relevant data to the end user from the environment presented in Fig. 2 and based on the developed taxonomy shown in Fig. 3, the elements of the conceptual system architecture are defined. The system architecture is based on IoT communication technology, associated sensors and devices, and an application solution is shown in more detail in Fig. 4.

Collecting and processing data in IoT environment can be presented by datasets, and all with the aim of defining accurate and real-time information to the end user [24, 25]. Based on the above, the elements presented in the conceptual architecture are presented through four layers. These layers are also described and presented through taxonomy. The layer of Perception and Actuator presents all the sensors, actuators, and devices that are used in the smart home environment. Depending on the user requirements, these sensors, actuators, and devices can be added and defined according to the user needs. Connecting and communication of the sensors,

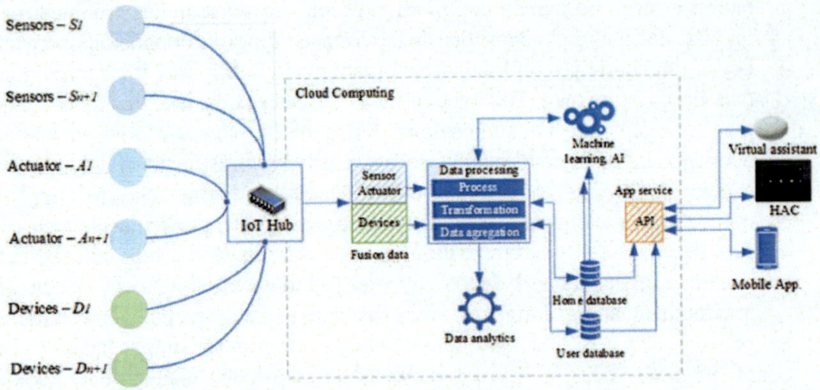

Fig. 4 Conceptual architecture of the system for the provision of information to visually impaired persons in a smart home

devices, and actuators is made possible through the communication technologies such as WiFi, Zigbee, Z-Wave, and similar. In the Preprocessing and Middleware layer (Cloud Computing) data collection and fusion are performed, as well as the processing according to user requirements. The user requirements are defined according to the type of disability and needs that the user has for the purpose of performing daily household needs and monitoring health. In Cloud Computing environment there is a user base and the base of all sensors, actuators, and devices and their characteristics.

Through M2H communication the user receives customized information by means of application on their mobile device or through the home system (virtual assistant). The applications used by the user, virtual assistant, or the smart home system are located in the Application layer. The design of the mobile application and the home application must include elements of universal design related to the accessibility of information, reduced physical effort in use, ensure equal method of usage for all types of users, flexibility in usage, simple and intuitive use and visible information. Management and communication of the user with the network of sensors, actuators, and devices in a smart home is easiest with a virtual assistant.

The goal of applying this conceptual architecture is to create a smart environment and create a user knowledge base based on everyday need and activities. The activities and user needs can be monitored by the sensors, actuators, and devices located in the user's environment. A communication network that would connect all sensors, actuators, and devices and that would enhance the user experience by using the virtual assistant is the autonomous network. The use of artificial intelligence in the smart home environment and the operation of the autonomous network would enable: a safer environment for the user, user interaction with the doctor (monitoring user health), connecting with other stakeholders, etc.

5 Testing the Effectiveness of Virtual Assistants on the Example of a Smart Home

Google Home and Amazon Alexa devices were tested to verify the operation of the virtual assistant in the smart home environment for the visually impaired persons. Testing was done in laboratory environment using the Arduino IoT development environment and sensors.

Monitoring of sensor operation and data collection is presented in Thingspeak Cloud application presented in Fig. 5. Figure 5 shows the collection of data in certain intervals.

Of the technical characteristics of sensors and modules, the following have been used:

- Humidity and temperature sensor – DHT11
- Color sensor – TCS230
- Motion sensor – HC-SR501
- Pulse – EN-11574
- Vibration – SW-420
- Module for connecting to WiFi network – ESP32

All sensors and modules are connected to Arduino Mega plate and virtual assistants Google Home and Amazon Alexa are used for their management.

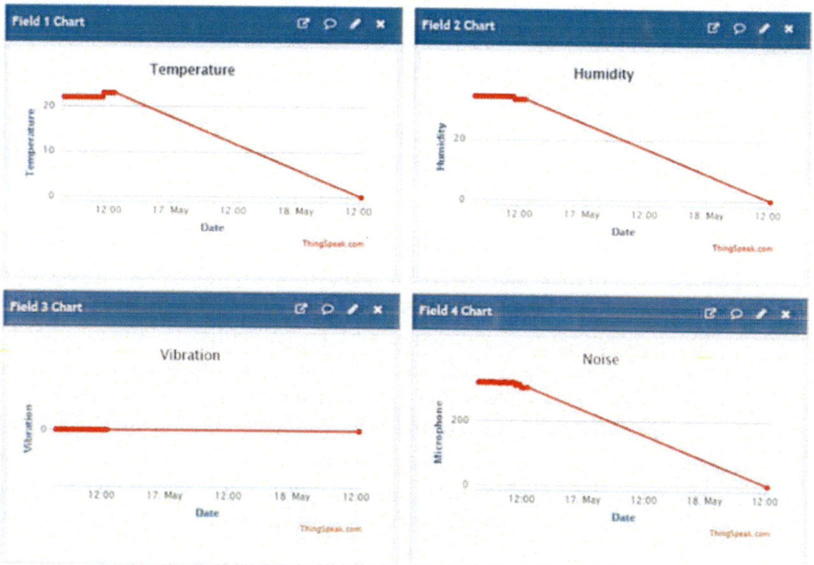

Fig. 5 Collection of data in Thingspeak environment

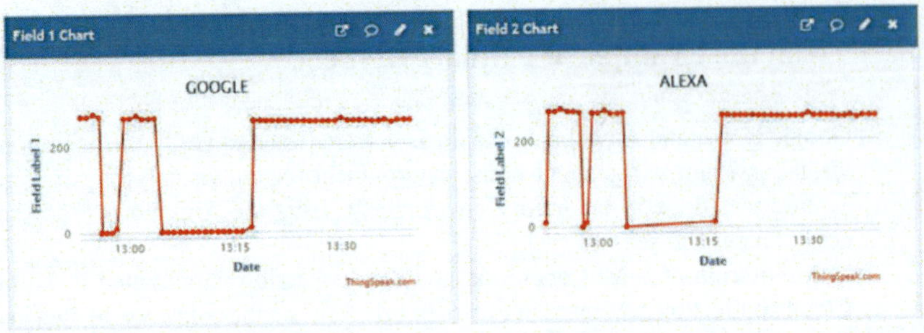

Fig. 6 Operation of virtual assistant

Measurements have also shown that Amazon Alexa has faster system response when making request for the start and management of the sensor as presented in Fig. 6.

The efficiency of the virtual assistant operation is important from the aspect of providing information to the user of the smart home, mostly because of the type of disability the user has. Equipping of the user with several devices for managing the sensors, actuators, and devices in the home can lead to discomfort in using, and it can also reduce the quality of life in the user's home environment. The lack of tested virtual assistants is the language support. The virtual assistant used for this purpose represents assistive technology that can raise the level of the quality of life and make everyday life easier for the user.

6 Conclusion

Following the trends and the development of assistive technologies for the help of visually impaired persons we can say today that it is not just a device, but a technology, service, and a system. Systems and services based on IoT technology can raise the level of the quality of life thus raising the level of mobility and inclusion. Designing and modeling need to be considered according to the user needs and on the basis of that, create a system customized for each user.

The possibilities provided today by information and communication networks and technologies such as artificial intelligence, machine learning, the autonomous networks can create a faster and more efficient system for the user. Based on the proposed conceptual architecture and the taxonomy, it is possible to create a smart home for the disabled persons depending on the type and degree of disability. Based on the defined elements of architecture it is possible to give the user all the information from the home environment and the information about the user health (pulse, temperature, blood pressure, sugar, etc.). Management of sensors, actuators, and devices in the home is possible by speech using the virtual assistant which enhances the user atmosphere in their homes.

References

1. Benjak, T. (2021). *Izvješće o osobama s invaliditetom u Republici Hrvatskoj.*
2. Rocha, A. P., Ketsmur, M., Almeida, N., & Teixeira, A. (2021). An accessible smart home based on integrated multimodal interaction. *Sensors, 21*, 5464. https://doi.org/10.3390/s21165464
3. Cvitić, I., Peraković, D., Periša, M., & Gupta, B. (2021). Ensemble machine learning approach for classification of IoT devices in smart home. *International Journal of Machine Learning and Cybernetics, 12*, 3179–3202. https://doi.org/10.1007/s13042-020-01241-0
4. Kepuska, V., & Bohouta, G. (2018). Next-generation of virtual personal assistants (Microsoft Cortana, Apple Siri, Amazon Alexa and Google Home). In *2018 IEEE 8th Annual Computing and Communication Workshop and Conference (CCWC)* (pp. 99–103). IEEE. https://doi.org/10.1109/CCWC.2018.8301638
5. Alexakis, G., Panagiotakis, S., Fragkakis, A., Markakis, E., & Vassilakis, K. (2019). Control of smart home operations using natural language processing, voice recognition and Iot technologies in a multi-tier architecture. *Designs, 3*, 1–18. https://doi.org/10.3390/designs3030032
6. W3 Organisation. *Synchronization Accessibility User Requirements (SAUR).*
7. Abdolrahmani, A., Kuber, R., & Branham, S. M. (2018). Siri talks at you: An empirical investigation of voice-activated personal assistant (VAPA) usage by individuals who are blind. In *ASSETS 2018 – Proceedings of the 20th International ACM SIGACCESS Conference on Computers and Accessibility* (pp. 249–258). https://doi.org/10.1145/3234695.3236344
8. Caporusso, N., Elleman, N., & Cho, S. S. (2020). Interface digital twins: Rendering physical devices accessible to people who are blind. In *2020 43rd International Convention on Information, Communication and Electronic Technology, MIPRO 2020 – Proceedings* (pp. 982–986). https://doi.org/10.23919/MIPRO48935.2020.9245373
9. Rizvi, S., Sohail, I., Saleem, M.M., Irtaza, A., Zafar, M., & Syed, M. (2018). A smart home appliances power management system for handicapped, elder and blind people. In *2018 4th International Conference on Computer and Information Sciences: Revolutionising Digital Landscape for Sustainable Smart Society, ICCOINS 2018 – Proceedings* (pp. 2–5). https://doi.org/10.1109/ICCOINS.2018.8510595
10. Triyono, L., Yudantoro, T. R., Sukamto, S., & Hestinigsih, I. (2021). VeRO: Smart home assistant for blind with voice recognition. *IOP Conference Series: Materials Science and Engineering, 1108*, 012016. https://doi.org/10.1088/1757-899x/1108/1/012016
11. Garg, R., & Cui, H. (2022). Social contexts, agency, and conflicts: Exploring critical aspects of design for future smart home technologies. *ACM Transactions on Computer-Human Interaction, 29*, 1–30. https://doi.org/10.1145/3485058
12. Bisio, I., Garibotto, C., Lavagetto, F., & Sciarrone, A. (2019). When eHealth meets IoT: A smart wireless system for post-stroke home rehabilitation. *IEEE Wireless Communications, 26*, 24–29. https://doi.org/10.1109/MWC.001.1900125
13. Boswarthick, D., Compans, S., Deschanel, M., Frost, L., Leadbeater, A., Lopez, D., Marchese, P., Owen, G., Paine, K., Toche, C., Ward, M., Holley, K., & Khun-jush, J. (2021). *ETSI Technology Radar, White Paper*. ETSI.
14. Donnici, R., Coronato, A., & Naeem, M. (2021). A self-learning autonomous and intelligent system for the reduction of medication errors in home treatments. In *Intelligent Environments 2021: Workshop Proceedings of the 17th International Conference on Intelligent Environments* (pp. 156–166). https://doi.org/10.3233/AISE210093.
15. Aski, V. J., Dhaka, V. S., Kumar, S., Verma, S., & Rawat, D. B. (2022). Advances on networked eHealth information access and sharing: Status, challenges and prospects. *Computer Networks, 204*, 108687. https://doi.org/10.1016/j.comnet.2021.108687
16. Kunnappilly, A. (2021). *Modeling and formal analysis of e-health systems*. Mälardalen University.
17. Peraković, D., Periša, M., Cvitić, I., & Zorić, P. (2019). Internet of things concept for informing visually impaired persons in smart factory environments. In L. Knapčíková & M. Balog

(Eds.), *Industry 4.0: Trends in management of intelligent manufacturing systems* (pp. 69–86). Springer. https://doi.org/10.1007/978-3-030-14011-3_7
18. Kim, M. J., Cho, M. E., & Jun, H. J. (2020). Developing design solutions for smart homes through user-centered scenarios. *Frontiers in Psychology, 11*, 1–12. https://doi.org/10.3389/fpsyg.2020.00335
19. Leporini, B., & Buzzi, M. (2018). Home automation for an independent living. In *Proceedings of the 15th International Web for All Conference* (pp. 1–9). ACM. https://doi.org/10.1145/3192714.3192823
20. TM Forum: Autonomous network technical architecture. (2021).
21. Kerravala, Z. (2019). *Digital transformation requires autonomous network*. ZK Research: A Division of Kerravala Consulting.
22. ETSI: Autonomous networks, supporting tomorrow's ICT business. (2020).
23. TM Forum: Autonomous network levels evaluation methodology. (2021).
24. Periša, M., Marković, G., Kolarovszki, P., & Madleňák, R. (2019). Proposal of a conceptual architecture system for informing the user in the IoT environment. *Promet – Traffic & Transportation, 31*, 37–47. https://doi.org/10.7307/ptt.v31i1.2677
25. Periša, M., Peraković, D., Cvitić, I., & Krstić, M. (2021). Innovative ecosystem for informing visual impaired person in smart shopping environment: InnIoTShop. *Wireless Networks, 5*, 469–479. https://doi.org/10.1007/s11276-021-02591-5

Comparison of the Principles of Head-Up Display Technologies for the Implementation of Augmented Reality in the Automotive Industry: A Study

Jozef Husár, Stella Hrehova, Lucia Knapčíková, and Justyna Trojanowska

1 Introduction

In today's world, we encounter the term augmented reality (AR) on almost every corner, although many times we do not even know that it is AR. Augmented reality is an enhanced version of the real world, dominated by digital visual objects, sound or other sensory stimuli that are shown to us through the latest technologies. This is a growing trend among companies dealing mainly with mobile computers and business applications [1]. When augmented reality is used by different users, data is collected and used for analysis. This can improve the specific real-world features and increase its understanding. The gained knowledge can be used then for real-world applications. These large-scale analyses can help companies make decisions and gain an overview of consumer habits within other companies as well [2, 3]. Augmented reality is an advanced technology that in today's world makes it easier for its users to interact with reality by adding virtual objects to it. This is the most natural way to connect with the digital world. Try to imagine looking at this as one optical view that provides information to explain everything and using AR to show us that many things in this world are not that complicated. Augmented reality is closely linked to mixed reality, where until recently they were just fiction, until they have recently become a science-based reality. The task of this reality is not

J. Husár (✉) · S. Hrehova · L. Knapčíková
Technical University of Košice, Faculty of Manufacturing Technologies with a seat in Prešov, Department of Industrial Engineering and Informatics, Prešov, Slovak Republic
e-mail: jozef.husar@tuke.sk; stella.hrehova@tuke.sk; lucia.knapcikova@tuke.sk

J. Trojanowska
Faculty of Mechanical Engineering, Department of Production Engineering, Poznan University of Technology, Poznan, Poland
e-mail: justyna.trojanowska@put.poznan.pl

L. Knapčíková, D. Peraković (eds.), *7th EAI International Conference on Management of Manufacturing Systems*, EAI/Springer Innovations in Communication and Computing, https://doi.org/10.1007/978-3-031-22719-6_2

to remove the object from the real world and move it to a completely different place, but on the contrary, the world in which we are improving with various 3D virtual objects that can improve our imagination for a given thing [4]. Recently, there have been opinions that the implementation of AR is needed, especially in the automotive industry, which is also the focus of this chapter. One of the biggest advantages of augmented reality, in particular, is that it allows us to have a more interactive and richer interaction between man and computer, which is mediated by human interaction as a conventional interface, which can be projected into more specific useful applications [5]. An important advantage of AR is that it expands the functions and capabilities of computers as well as other consumer electronic devices, such as game consoles, smartphones and wearable technologies. In this way, AR is constantly improving the user environment of these devices and enriching the user experience more and more. Compared to virtual reality, another specific advantage of augmented reality is that it does not replace the natural environment with simulated ones, but only adds other 3D objects, thus inserting parts of the digital world into the perception of the real world [6–8]. Augmented reality provides more freedom for users and more options for traders because a display that is mounted on the head may not be provided. Augmented reality has much greater market potential than virtual reality, so its development and growth is advancing at a faster pace as large companies and brands begin to implement it. AR is not so much affected by the limitations of the devices on which it is used. Nevertheless, there are still many demands on the creation of objects that work with very high resolution and are in fact indistinguishable from realistic objects [9].

1.1 AR in Automotive Industry

The automotive industry is interested in the development of AR and its applications in several ways. Since the end of the twentieth century, automotive developers and original equipment manufacturers (OEMs) have been addressing the benefits of AR through large joint projects [10]. During the implementation of the projects, it was an effort to use AR throughout the product life cycle, from the beginning of the car concept to the final customer assistance services. Although developments have helped to integrate AR into the automotive industry, its use is sometimes still limited due to technical shortcomings [11].

Devices for using augmented reality use the position of the user's angle of view with orientation and position coordinates and subsequently project the information in a certain way. For example, on the screen of a mobile device with a real background or in the lenses of AR glasses. The development of mobile devices has thus allowed great growth in technologies that provide various tools and platforms for AR applications in these devices [12]. A good example is when Volkswagen has created an augmented reality digital system for digital production line testing. This model uses a 3D assembly model of the structure, which combines the image with data with the help of a camera that captures the structure during its production. It compares the target state with the actual state, which coincide in millimetres [13]. It is also worth mentioning the good usability of spatial augmented reality (SAR), especially due to its good usability. For example, to overcome some technologies such as ergonomic failure of the AR system when the systems are used for longer periods. In fact, AR differs from SAR in that it does not require the use of glasses and screens because the information is projected directly in daylight. There are many areas for implementing AR in the automotive industry. First and foremost, it's factory planning. Here, AR allows us to make adjustments and optimizations without interfering with the real line [14]. These interventions are performed in a digital model, then simulated and after obtaining the values, the inputs to the real line can be approached. Another area is the concept and design of cars, where we can implement interventions in the design and model of the car before its production. We can display the car in a scale of 1: 1 and with the help of mixed reality we monitor and adjust the elements of the car in the digital version before its production in the pre-production stage. Another little-presented area is the use of AR in the production of cars for the assembly and disassembly of cars. During production on the line, the individual types of cars come in a precisely determined order (e.g. sedan, station wagon, SUV, hatchback). The employee on the line must control which tasks he has to perform on which position and in what order according to the technological procedure. AR can help here, where after scanning the object, it points to the employee digital objects placed in the appropriate places [15–19]. AR can also be used in sales, where, similarly to car design, the client can assemble the car exactly according to his ideas. It can make changes to the interior, colour and design on a digital model and then project it into free space. With AR, we can improve the customer experience and thus show him a better idea with the help of interactions that take place between the real environment and virtual elements. AR as a user manual can provide us with an interesting alternative to the classic paper manual that you receive with each vehicle purchased. In such manuals, you will find various instructions, diagrams and pictorial forms that apply to your vehicle. By delivering these steps to the real vehicle, the AR can avoid possible ambiguities. If the user does not know exactly whether he is looking at the right part of the vehicle, the AR will direct him to the exact place [20]. However, if the interactive manual is not sufficient and the problem is beyond the user's ability, an online video stream of the AR device led by a remote service technician can help. The last area that we give priority to in this chapter is the use of AR as a driving assistant. When driving, the driver must focus on the perception of the environment, as this is a key element of the driving function. We can use augmented reality to improve perception in several ways [21]. Augmented reality obscures the driver's field of vision and shows the trajectory during the manoeuvre. Augmented reality can choose the right trajectory that can select the right path for the driver through navigation. In another case, augmented reality could help the driver reduce his subconscious efforts, because compared to today's navigation and current assistants, the driver would not have to focus on the perception of the surroundings. Augmented reality is intended to help the driver, in particular, increase attention in situations such as junctions and crossings and thus avoid the various dangers associated with other

Fig. 1 AR as an assistant in driving a vehicle

road users [22, 23]. The AR is intended to help the driver improve his perception of the surroundings if conditions are aggravated by weather or other reasons. In fog conditions, road edges should be highlighted and vehicles that are not clearly visible because they overlap with a building or other vehicle will be displayed to the driver using AR. Ultimately, by developing an autonomous vehicle, we can help develop entertainment and information applications, such as a tourist guide where augmented reality will help us a lot, in providing this information [24]. While the vehicle is precisely located using a limited VSLAM camera, rough 3D city mode and standard GPS (top left), AR is used to display the route to follow (top right) such as dangerous intersections, pedestrian crossings (bottom left), (top right) or tourist information, such as a hotel (bottom right) (Fig. 1).

2 Research

In this section, we give priority to augmented reality as a management assistant. There are many ways to project the required information into the field of view. With the advent of digital displays, much more information for the driver began to be projected than the classic indicators on the dashboard. For example, the driver of the car may set the navigation projection as a priority and the speed and condition information of the car is projected on the sides of the on-board unit. This opens up the possibility of augmented reality and projection of selected parameters on the head-up display [25].

If we talk about the head-up display, we are actually talking about a digital-transparent image, which is projected on the windshield of the vehicle. The role of the HUD is to replace devices that are located on the dashboard, such as speed indicator or navigation. The use of HUD is expected to increase safety by displaying optical elements in front of the driver's eyes. This helps the driver increase his focus to the direction of travel without having to look at different screens while seeing the same information that is projected on the HUD. Head-up displays in automobiles generally use the projection of light on the windshield and the elements are thus displayed as reflections. HUD creates a continuous superposition between HUD elements and environmental events. Placing the element in the correct position is very important when using the HUD, because the good readability of the information is affected by the background. On the head-up display, which is connected to augmented reality (AR-HUD), the parts overlap with the external optical world by projection on the windshield. With AR-HUD technology, drivers have better access to the information they need without having to look away from the road [26, 27].

The main task of the AR-HUD is to create safer vehicles, but there are still some concerns about driver distraction. When driving a vehicle that is equipped with a head-up display with augmented reality, the driver watches the information on the display, but also has to pay enough attention to the road and such a phenomenon could result in inattention. Uniti supplies its vehicles with a wider range of applications on the market compared to the competition, as the HUD is slowly replacing the classic instrument panel. All design challenges or design opportunities speak of all displays of information that are necessary for the driver in a clear and clear manner. If we want to talk about HUD implementations, we must specify the basic requirements for AR-HUD [28].

Technological Implementation

The technological processes of ordinary HUD display work on the reflection of light. The projection or 2D display unit has its place in a glass pane, which is placed at an angle of 40°–45°, at which light is reflected and thus creates a virtual image visible to the observer. The observer observes the virtual image at a distance that is adequate to the light path between the combiner and the eyes. It is the distance from the source to the eye. Optical elements such as lenses and mirrors can extend the distance of the light path and help keep the dimensions of the entire optical system as low as possible, which can limit the space of the dashboard. These elements limit the distance and size of the created image.

Image Dimensions

The field of view that the driver perceives when driving measures approximately 67° × 20°. Projecting an HUD image takes up approximately 6° × 2° of space. The head-up display is limited by the dashboard space and is related to the image distance. Increasing the image distance reduces its quality and perceived size. With the available space on the dashboard, the size of the image also increases linearly.

Image Location
The head-up display has its place on the dashboard of the car, where it is located 5° below the driver's line of sight. Many attempts have been made for the most suitable location of the HUD based on different driver preferences and recognition. The researchers narrowed the area they decided to test to $20°-10°$ around the driver's field of vision in order to reduce the range of findings. Many drivers who use navigation devices have them located above the console centre. According to the researchers, this site uses high monitoring power, but the placement of the HUD at this site was excluded.

Image Distance
At a distance of approximately 2–2.5 m, the image is taken from the driver's eyes. With this distance, the accommodation of the eye decreases, which is actually the focusing between the light and the display. Many studies have been conducted to find the best viewing distance, where it has been found that a distance of 2–3 m is the most optimal room distance. When tested at a distance of 1–5 m, it was shown that older drivers experienced an increase in stimulus recognition times closer than 2.5 m. In this research, the range was limited to a few meters, so these participants could not test the perception at greater distances. Suppliers and car manufacturers say that a better distance than a state-of-the-art HUD is needed for the HUD to function and perceive better.

Contents of Head-Up Displays
The main task of head-up displays is mainly to offer information and driving data. The content of the HUD is based on the functions offered by the car and the data made available by the manufacturers. We most often encounter indicators of speed, speed limit or other restrictions, a warning about the procedure, navigation route or vehicle status, where we are shown, for example, the fuel status [29].

Nowadays, we can watch more and more content there, which is no longer completely related to driving, and thus contains, for example, a phone book, calls or music control functions. There are also external head-up display devices, such as the Garmin, Pioneer's NavGate HUD or Navdy, which can be placed under the windshield. Today, almost everyone is a smartphone holder who, thanks to the Navier HUD navigation or HUDway application, can also use it as a HUD located under the windshield. External HUDs are smaller in size than built-in HUDs, but have a more diverse content. Navdy HUD is one of the most popular because it allows a lot of information that is associated with smartphones such as messaging, phone call control, music control or appointment scheduling.

3 Methodology

In this part of the chapter we deal with three types of HUD. The first is direct projection on the windshield, as an example we chose the head-up display from Mercedes, which is a pioneer in the research and application of interactive head-up displays. The second type is projection on the reflecting surface and an example is the HUD from Citroen, where the projection display automatically pops up after start-up. The third alternative for comparison is the use of mobile applications and its use as a HUD.

3.1 Mercedes HUD

The head UP display from Mercedes has an opening angle of 10° horizontally and 5° vertically, so the image is displayed virtually at a distance of 10 m. Such a display area can be compared to a monitor with a diagonal of 77 inches. AR-HUD provides a wealth of augmented reality content for driving assistance systems and navigation information. For the driver, the image blends in perfectly with the surroundings, so there is no distraction. The image is created in high resolution with 1.3 million pixels. In the S-Class, Mercedes-Benz was the first to use it to create images on the head-up display.

Head-up display features:

- Full-colour display module, backlit by high-power LEDs.
- Special windscreen with built-in wedge-shaped sheet of laminated foil.
- Lens and mirror system.
- Light sensor in the area of the upper edge of the roof to adjust the brightness.
- Display menu settings.

On the head-up display offered by Mercedes, we can monitor functions such as speed, navigation, assistants or the time of arrival at the destination. There are also large directional arrows that change their position from right to left, depending on where we want to turn. Figure 2 shows an arrow that shows the direction to the left, as we can see from the navigation. This feature helps drivers to orient themselves more accurately on the road.

With lower visibility, the head-up display helps the driver increase the visibility of the curb, for example in a curve by highlighting lines in red. This allows the driver to see the curb better, which may not be visible enough (Fig. 3).

The head-up display, together with the adaptive cruise control, measures the position of the vehicle in front of you, which helps to maintain a sufficient distance. The distance of the vehicle in front of you is displayed in green and the exact distance can also be determined using the functions. The head-up display shows the exact navigation of the route, where after arriving at the destination, the destination point is projected on the windshield, which has the shape of a flag and signals the destination of the route (Fig. 4).

Fig. 2 Mercedes HUD

Fig. 3 HUD roadside warning and vehicle location

Fig. 4 HUD Mercedes – defining the route destination

Fig. 5 Citroen HUD – sliding screen and height setting

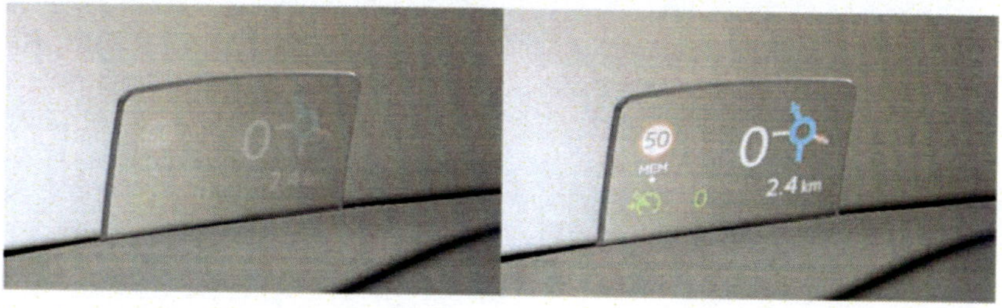

Fig. 6 Night and day mode settings

3.2 *Citroën Head-Up Vision*

Thanks to head-up vision, Citroën offers the user an overview of basic driving information. The system projects the data onto a pull-out screen located in the driver's field of vision. The driver of the vehicle is thus constantly informed about the driving speed, traffic sign data, cruise control information, navigation instructions or warnings of a possible collision (Fig. 5).

When the vehicle is started, a screen is displayed that projects driving data. The driver will thus have a view of the speed, the limiter and the navigation, which is linked to a coloured direction indicator on which the directional arrows intersect, or to the correct exit from the roundabout.

UP/DOWN – using the UP/DOWN buttons, the user can correct the height of the Head UP indicators.

SUN/MOON – using the SUN/MOON buttons, the user can correct the brightness of the projection, for better visibility during the day or night (Fig. 6).

3.3 *Head-Up Display Using a Smartphone*

In order for drivers to have a head-up display in their older cars, they don't really need much. All you need is the right smartphone application that turns your phone's

Fig. 7 NAVIER HUD 3

screen into a mirror image, where you can adjust the brightness so that the reflection on the windshield is clearly visible during the day. It is necessary to place the smartphone in the right place on the dashboard so that it does not block the driver's view and its reflection is well projected. There are many applications available on the Internet that offer their head up display features for free. In this section, we compare two applications – NAVIER HUD 3 and HUDWAY GO

3.3.1 NAVIER HUD 3

NAVIER HUD 3 is ranked among the best HUD display applications that is available for Android for free. For the application to work properly, all you need is a 3G connection. The application is properly designed and so when projected on the windshield, a crystal-clear reflection is seen. The orientation and brightness of the screen automatically adjusts while driving, so the driver has no worries about setting up the application while driving, the route can also be set by voice control. The application has an OBD2 function, which means that it uses a reader to read data from the vehicle, such as speed, fuel level, vehicle errors and other technical features (Fig. 7).

3.3.2 HUDWAY Go

The HUDWAY GO application is available for both IOS and ANDROID users. An important element for the application is navigation, which offers several options, such as voice control. Maps are also available offline. The graphics and brightness on the windshield are compatible and adjust automatically, but under certain conditions the image is not clearly visible enough. The application displays real

Fig. 8 HUDWAY GO

driving data such as average speed, time to destination or travelled distance. After turning off the engine, the application evaluates the driving data it provides to the user. Another feature that the application offers is the monitoring of fuel costs (Fig. 8).

4 Results

For comparison, we chose from three different types of head-up displays. We were the first to introduce a head-up display from Mercedes. Mercedes relies primarily on its quality, as evidenced by the design and display of the head-up display. It is connected to a 12.8-inch touchscreen, which is located in the centre of the dashboard. With this screen, it is possible to precisely configure and set different types of views of head-up display. The image, which is projected into the driver's field of view, captivates with its graphic design and also very good visibility even on a clear day. The driver no longer has to take his eyes off the road to check the speed, because it is displayed using the head-up display together with the display of the maximum permitted speed of the road on which the car is currently travelling. The display also shows icons of activated assistants, such as adaptive cruise control, where when the safe distance is exceeded, the driver is shown a red barrier or lane assist, which scans lanes and draws road edges in red lines. With the help of navigation, which is projected in the driver's field of vision, no one gets lost anywhere. The large directional arrows show the exact direction and number of meters before the vehicle enters lanes or intersections, which makes it easier for the driver to orient himself even in more frequented places. The display also shows the

Table 1 Comparison of mobile applications

Parameters	NAVIER HUD 3	HUDWAY GO
Price	Free	Free
Premium option	9.99$	8.99$
Android	Yes	Yes
IOS	Yes	Yes
Maps/navigation	Yes	Yes
3D view	Yes	Yes
Voice alerts	No	Yes
Voice control	Yes	Yes
Current speed	Yes	Yes
Driving time	Yes	Yes
Distance travelled	No	Yes
Driving analysis	No	Yes
Consumption calculation	No	Yes
OBD2	Yes	No
Visibility during the day	Good	Weak

number of kilometres to the destination and the destination itself, which is finally shown by a large symbol.

Although Citroen is in a lower price range than Mercedes, the developers have come up with a unique head-up display. The image is no longer projected on the road in front of the driver, but is displayed on a transparent sliding screen that is still in the driver's field of vision. When the head-up display is switched on, the screen on which the driving data is projected is displayed. The display shows a simpler image on which, however, speed or navigation is clearly visible. The navigation display is in a lower graphic quality, but it is still transparent and the roundabouts are very well drawn. The function display is height-adjustable as well as the brightness. During a sunny day, there may be less visibility because it is a transparent image. The functions as well as the processing of the head-up display itself, which Citroen offers, have received very good reviews in its price category.

Heads of displays found in cars can still be seen on the road in small numbers, as they are elements of new cars. However, owners of older types of cars keep up with the times and use applications that can at least partially replace the built-in head-up displays. We have selected 2 applications that are freely available for the market and completely free. NAVIER HUD 3 and HUDWAY GO are applications that can project an image from a smartphone onto the windshield and can replace the built-in head-up display (Table 1).

Application NAVIER HUD 3 offers better picture quality compared to HUDWAY GO and also better visibility on a clear day. HUDWAY GO, on the other hand, offers more options and functions that can be projected on the windshield, and the application is also easier to understand. NAVIER HUD 3 has OBD2, which means that the application can project the driver data on the vehicle, which it obtains from the vehicle using the OBD2 unit. These are data such as fuel level, tire pressure or vehicle errors.

Table 2 Comparison head-up displays

Parameters	Mercedes	Citroen	Applications
Price	Higher equipment	Higher equipment	Free
Image quality	Excellent	Very good	Good
Visibility during the day	Excellent	Very good	Good
Language	SK	SK	ENG
Function richness	1.	3.	2.
Voice alerts	Yes	No	Yes
Driving analysis	Yes	Yes	Yes

5 Conclusion

In this article, we pointed out three completely different types of head-up displays. Each display was something specific. Mercedes projects the image on the windshield, Citroen on the slide-out screen and the head-up display using the application is projected on the windshield using a smartphone. Mercedes stands out because of the sophisticated navigation, which shows directional arrows in space, Citroen combines simplicity with all the necessary information, and the head-up display with the application has limited functions, but it can work in any car (Table 2).

In conclusion, we can state that augmented reality is becoming a major milestone in the production and use of cars, where it already occupies a huge part. It helps not only car manufacturers but also our users who use these products associated with it. In the future, it can be assumed that with the development of autonomous vehicles, the implementation of AR will also develop, especially in the field of augmented reality.

Acknowledgements This work was supported by the projects VEGA 1/0268/22 and KEGA 038TUKE-4/2022 granted by the Ministry of Education, Science, Research and Sport of the Slovak Republic.

References

1. Buń, P., Trojanowska, J., & Rewers, P. (2019). VR and AR in lean manufacturing classes. In J. Trojanowska, O. Ciszak, J. Machado, & I. Pavlenko (Eds.), *Advances in manufacturing II. MANUFACTURING 2019* (Lecture Notes in Mechanical Engineering). Springer. https://doi.org/10.1007/978-3-030-18715-6_29
2. Dunleavy, M. (2014). Design principles for augmented reality learning. *Tech Trends, 58*, 28–34. https://doi.org/10.1007/s11528-013-0717-2
3. Kascak, J., et al. (2019). Implementation of augmented reality into the training and educational process in order to support spatial perception in technical documentation. In *IEEE 6th international conference on industrial engineering and applications (ICIEA)* (pp. 583–587). IEEE. https://doi.org/10.1109/IEA.2019.8715120

4. Stadler, S., Cornet, H., Mazeas, D., Chardonnet, J.-R., & Frenkler, F. (2020). ImPro: Immersive prototyping in virtual environments for industrial designers. In *Proceedings of the design society: DESIGN conference* (pp. 1375–1384). Cambridge University Press. https://doi.org/10.1017/dsd.2020.81
5. Trojanowski, P., & Trojanowska, J. (2021). Reliability of road transport means as a factor affecting the risk of failure – The transport problem case study. In V. Ivanov, J. Trojanowska, I. Pavlenko, J. Zajac, & Perakovi'c, D. (Eds.), *Advances in design, simulation and manufacturing IV. DSMIE 2021* (Lecture Notes in Mechanical Engineering) (pp. 253–261). Springer. https://doi.org/10.1007/978-3-030-77719-7_26
6. Židek, K., Maxim, V., Pitel, J., & Hosovsky, A. (2016). Embedded vision equipment of industrial robot for inline detection of product errors by clustering-classification algorithms. *International Journal of Advanced Robotic Systems, 13*, 1–10. https://doi.org/10.1177/1729881416664901
7. Himperich, F. (2007). *Applications in augmented reality in the automotive industry* (pp. 1–21). Fachgebiet Augmented Reality, Department of Informatics.
8. Židek, K., et al. (2020). Digital twin of experimental smart manufacturing assembly system for industry 4.0 concept. *Sustainability, 12*(9), 3658. https://doi.org/10.3390/su12093658
9. Kujawińska, A., & Diering, M. (2021). The impact of the organization of the visual inspection process on its effectiveness. *The International Journal of Advanced Manufacturing Technology.*, JAMT-D-20-02941R1, *112*, 1295–1306. https://doi.org/10.1007/s00170-020-06543-9
10. Varela, M. L. R., Putnik, G. D., Manupati, V. K., et al. (2021). Integrated process planning and scheduling in networked manufacturing systems for I4.0: A review and framework proposal. *Wireless Networks, 27*, 1587–1599. https://doi.org/10.1007/s11276-019-02082-8
11. Gay-Bellile, V., Bourgeois, S., & Tamaazousti, M. (2015). Applications of augmented reality for the automotive industry. In W. Barfield (Ed.), *Fundamentals of wearable computers and augmented reality* (2nd ed., pp. 433–456). CRC Press.
12. Antosz, K., & Stadnicka, D. (2018). An intelligent system supporting a forklifts maintenance process. In A. Burduk & D. Mazurkiewicz (Eds.), *Intelligent systems in production engineering and maintenance – ISPEM 2017. ISPEM 2017* (Advances in Intelligent Systems and Computing) (Vol. 637). Springer. https://doi.org/10.1007/978-3-319-64465-3_2
13. Tokorova, M., et al. (2020). Augmented reality as a support tool in machining process. *Tem Journal, 9*(1), 407–411. https://doi.org/10.18421/TEM91-56
14. Sharma, R., & Molineros, J. (1997). Computer vision-based augmented reality for guiding manual assembly. *Presence: Teleoperators and Virtual Environments, 6*(3), 292–317. https://doi.org/10.1162/pres.1997.6.3.292
15. Trojanowski, P. (2020). Comparative analysis of the impact of road infrastructure development on road safety–a case study. *Scientific Journals of the Maritime University of Szczecin, 63*(135), 23–28. https://doi.org/10.17402/436
16. Boboc, R. G., Gîrbacia, F., & Butilă, E. V. (2020). The application of augmented reality in the automotive industry: A systematic literature review. *Applied Sciences, 10*, 4259. https://doi.org/10.3390/app10124259
17. Gorski, F., Bun, P., & Stefanska, K. (2022). Use of virtual mirror in design of individualized orthopedic supplies. In J. Machado, F. Soares, J. Trojanowska, & E. Ottaviano (Eds.), *Innovations in mechanical engineering. ICIENG 2021* (Lecture Notes in Mechanical Engineering). Springer. https://doi.org/10.1007/978-3-030-79165-0_36
18. Demčák, J., Lishchenko, N., Pavlenko, I., Pitel', J., & Židek, K. (2022). The experimental SMART manufacturing system in SmartTechLab. In J. Trojanowska, A. Kujawińska, J. Machado, & I. Pavlenko (Eds.), *Advances in manufacturing III. MANUFACTURING 2022* (Lecture Notes in Mechanical Engineering). Springer. https://doi.org/10.1007/978-3-030-99310-8_18
19. Häuslschmid, R. (2018). *Extending head-up displays: Exploring the potential of large & 3D automotive windshield displays.* Dissertation, LMU München.https://doi.org/10.5282/edoc.22766

20. Kaščak, J., Husár, J., Knapčíková, L., Trojanowska, J., & Ivanov, V. (2022). Conceptual use of augmented reality in the maintenance of manufacturing facilities. In J. Trojanowska, A. Kujawińska, J. Machado, & I. Pavlenko (Eds.), *Advances in manufacturing III. MANUFACTURING 2022* (Lecture Notes in Mechanical Engineering). Springer. https://doi.org/10.1007/978-3-030-99310-8_19
21. Kawęcki, N., & Gola, A. (2022). Pick performance system as an IT support for order completing – A case study. In J. Trojanowska, A. Kujawińska, J. Machado, & I. Pavlenko (Eds.), *Advances in manufacturing III. MANUFACTURING 2022* (Lecture Notes in Mechanical Engineering). Springer. https://doi.org/10.1007/978-3-030-99310-8_9
22. Trojanowski, P., Trusz, A., & Stupin, B. (2022). Correlation between accidents on selected roads as fundamental for determining the safety level of road infrastructure. In V. Ivanov, J. Trojanowska, I. Pavlenko, E. Rauch, & D. Peraković (Eds.), *Advances in design, simulation and manufacturing V. DSMIE 2022* (Lecture Notes in Mechanical Engineering). Springer. https://doi.org/10.1007/978-3-031-06025-0_11
23. Menk, C., Eduard, J., & Reinhard, K. (2010). Evaluation of geometric registration methods for using spatial augmented reality in the automotive industry. In *Vision, modeling, and visualization workshop* (pp. 243–250). https://doi.org/10.2312/PE/VMV/VMV10/243-250
24. Saxena, D. (2022, November 10). *How to use your smartphone as heads-up display*. https://wethegeek.com/how-to-use-your-smartphone-as-heads-up-display. Accessed 16 June 2022.
25. Kluz, R., & Antosz, K. (2019). Simulation of flexible manufacturing systems as an element of education towards industry 4.0. In J. Trojanowska, O. Ciszak, J. Machado, & I. Pavlenko (Eds.), *Advances in manufacturing II. MANUFACTURING 2019* (Lecture Notes in Mechanical Engineering). Springer. https://doi.org/10.1007/978-3-030-18715-6_28
26. Sharma, L. (2019). Recent advancements of augmented reality in real time applications. *International Journal of Recent Technology and Engineering, 8*, 537–542. https://doi.org/10.35940/ijrte.B10100.0782S719
27. Lv, Z., Halawani, A., Feng, S., et al. (2015). Touch-less interactive augmented reality game on vision-based wearable device. *Personal and Ubiquitous Computing, 19*, 551–567. https://doi.org/10.1007/s00779-015-0844-1
28. Kłosowski, G., Gola, A., & Thibbotuwawa, A. (2018). Computational intelligence in control of AGV multimodal systems. *IFAC-PapersOnline, 51*(11), 1421–1427. https://doi.org/10.1016/j.ifacol.2018.08.315
29. Antosz, K., Pasko, L., & Gola, A. (2019). The use of intelligent systems to support the decision-making process in lean maintenance management. *IFAC PapersOnLine, 52*(10), 148–153. https://doi.org/10.1016/j.ifacol.2019.10.005

Polygonization of the Surface Digitized Using Helios2 Time-of-Flight Camera

Adrián Vodilka, Martin Pollák, and Marek Kočiško

1 Introduction

Time-of-flight (ToF) technology is based on the calculation of reflected flight time [1]. ToF technology is a technology used for decades to measure the distance between the sensor and the object. This technique is based on illuminating objects with a modulated light source and analysis of reflected light [2]. The illumination is performed in repeated pulses and near-infrared light is used. The phase shift between the pulse of light from the source and its reflection from the surface of the object is measured and used to calculate the distance. The most commonly used light source is a laser light or LED and an infrared light is used. The same basic principle ToF of calculating the flight time of reflected light is also used when measuring distances with laser distance gauges [3]. Using a ToF camera, it is possible to obtain a depth image, with a depth distance to the surface of a physical object in the captured scene being obtained for each point of the camera. The main applications of ToF implementation include robotic vision, 3D inspection, manipulation of physical objects, object sorting, palletizing and depalletizing, volume determination of physical objects, environmental sensing of autonomous vehicles, logistics, 3D digitization of objects, human interaction with machinery, and others [4].

Laser light sources or LED diodes with emitted light at a wavelength of 850 nm, which is invisible to the human eye, are used as a source of illumination. The optical sensor captures light at this specific wavelength and converts it into an electric current. The depth distance is measured for each point of the 2D plane within the camera's ToF coordination system, creating an image depth map and point cloud.

A. Vodilka (✉) · M. Pollák · M. Kočiško
Technical University of Kosice, Faculty of Manufacturing Technologies, Presov, Slovakia
e-mail: adrian.vodilka@tuke.sk; martin.pollak@tuke.sk; marek.kocisko@tuke.sk

L. Knapčíková, D. Peraković (eds.), *7th EAI International Conference on Management of Manufacturing Systems*, EAI/Springer Innovations in Communication and Computing, https://doi.org/10.1007/978-3-031-22719-6_3

By polygonizing a point cloud, it is possible to obtain a polygonal surface model of objects [5].

2 Point Cloud Acquisition Using Helios2 Time-of-Flight Camera

2.1 *Helios2 Time-of-Flight Camera*

The Helios2 time-of-flight camera is a 3D optical imaging device designed by LUCID Vision Labs Inc. in Canada, using a Sony DepthSense IMX556PLR CMOS sensor and reflected light time calculation digitization technique to calculate the distance between the depth of the device and the scanned surface of physical objects. The company's first ToF device was introduced in 2019 called the Helios ToF 3D camera, making it the first commercially available ToF industrial device using the innovative Sony DepthSense sensor. Later, the Helios Flex 3D ToF MIPI product was introduced, a module with lower dimensions working with the Nvidia Jetson TX2 computing device using a Flex cable. In 2020, the company presented the Helios2 IP67 ToF 3D camera with improved accuracy and quality of digitized data. The Helios2 camera is designed to work indoors, suitable for a low-window industrial environment. For outdoor use, it is necessary to use a ToF camera with illumination with a higher wavelength (e.g., 950 nm). Helios2 is a durable device with dimensions of 60 × 60 × 77.5 mm, weight 398 g, and thanks to IP67 certification is suitable for demanding industrial use, operating range −20 to 50° C without the use of cooling and fans and is designed for continuous operation. Helios2 camera technology is designed to perform more accurate digitization of the surface of objects compared to other devices using ToF technology, focusing on the quality of details, noise reduction, and filter of flying points. The device resolution is 640 × 480 px (0.3 MP) at a refresh rate of 30 fps in all operating modes. The device's optics allow digitization with a viewing angle in the range of 69° × 51°. The device is designed with respect to the GigE Vision v2.0 standard for communication within the GenICam 3D interface, while enabling communication via the GigE and M12 X-coded digital interfaces and the GPIO interface using an 8-pin M8 connector. The device can be powered using PoE + IEEE 802.3at or using GPIO with a voltage of 18–24 V. The average use of the device is 12 W with a maximum power of 30 W. Thanks to the five communication channels of the device, it is possible to use up to five Helios2 devices simultaneously without mutual interference [6–8] (Fig. 1).

Helios2 Accuracy and Precision It is important to distinguish between accuracy and precision when working with Helios2. Accuracy defines the difference between the actual position and the position within the digitized surface of the object. Precision defines the standard deviation of the acquired data. Helios2 allows digitization in 6 operating distance modes ranging from 0.3 to 8.33 m. Operating

Fig. 1 Helios2 time-of-flight camera

Table 1 Distance modes accuracy table of Helios2 [8]

Operating mode (mm)	Modulation frequency (MHz)	Accuracy
1250	120	±4 mm
3000	50	±10 mm
4000	37	±10 mm (+0.25% of depth)
5000	120 + 90	±4 mm (+0.1% of depth)
6000	25	±10 mm (+0.5% depth)
8300	90 + 72	±4 mm (+0.2% depth)

distance modes operate at different modulation frequencies and the parameter determines the maximum operating distance of scene objects to the device. A refresh rate of 30 FPS can be achieved within all operating distance modes. When selecting a suitable mode, it is necessary to take into account the required distance of digitizing the scene and the values of accuracy and precision, so that it is possible to digitize the scene in the highest quality of the obtained data. The accuracy and precision of the acquired data depend on the temperature of the camera, while the distance of the measured depth increases as the temperature increases. The Helios2 camera is calibrated to compensate for temperature changes in a sufficient range, without the need for self-calibration. The change caused by temperature cannot be completely eliminated by compensation, therefore the camera must be allowed at least 15 minutes to warm up before use [8] (Table 1).

As can be seen in the figure, the lowest depth deviation can be achieved in the range of 1–3 m. At the same time, it is possible to conclude that in operating modes with higher and dual frequency modulation, the accuracy and precision of the device are higher, namely 1250, 5000, and 8300 mm (Table 2) [9, 10].

ArenaSDK The Arena Software Development Kit (ArenaSDK) is a software package that provides an API programming interface for LUCID Vision Labs Inc. devices. Within the ArenaSDK, libraries and program examples are available in the C ++, C, .NET, and Python programming languages. The ArenaSDK includes a GUI graphical interface ArenaView based on the GenICam standard.

Table 2 Distance modes precision table of Helios2 [8]

Distance (m)	1250 mm mode	3000 mm mode	4000 mm mode	5000 mm mode	6000 mm mode	8300 mm mode
0.5	1.0	1.9	2.1	0.7	3.6	0.8
1	0.8	1.3	2.1	0.6	2.7	0.6
1.5	1.1	2.5	2.9	0.9	4.0	1.1
2	1.8	3.7	4.9	1.4	7.8	1.7
3		5.7	8.6	2.2	10.0	2.5
4			12.3	3.3	15.7	4.1
5				5.1	28.1	6.1
6					30.1	7.9
7						11.8
8						14.5

With ArenaView, it is possible to use and control the advanced features of the Helios2, while creating in-depth images or viewing a 3D point cloud in real time. With ArenaView it is possible to work in 2D and 3D display – 2D from the camera view and 3D point cloud display with the ability to change the view orientation in real time so it is possible to view the current digitized data or capture a snapshot. ArenaView exports Helios2 digitization data in 3D PLY and RAW format [11].

Using Helios2 within the ArenaView GUI is user-friendly. After the correct connection of the selected method of power supply and communication interface with the computer, the ArenaView software could be started, which enables the Helios2 device function and start the Stream. The ArenaView GUI launches the default view of the Helios2 device in the pixel format Coord3D_ABCY16 – four-channel XYZ point cloud data. Within the ArenaView user interface it is possible to change various parameters of the Helios2 device and GUI divided into three windows: Features, Properties, and Options [12].

By adjusting the features parameters, it is possible to select the correct settings for specific examples within the scene capture. The main options for adjusting the features parameters include Image Accumulation, Flying Pixel Filter, Exposure Time, Conversion Gain, Confidence Threshold, and Spatial Filter [11].

Helios2 Features and Options For the digitization of objects using the ToF technique, the most suitable objects are those whose surface is not very reflective or too absorbing light. At high levels of light absorption, gaps in the point clouds typical of black surfaces can occur because the ToF sensor does not detect absorbed light on the object's surface. For glossy surfaces, points in glossy areas can register in front of or behind the actual coordinate, thus reducing accuracy. Image Accumulation in the range of 0–32 images is used to improve the obtained data by translating multiple images into a common calculation of average values. A higher value of this parameter improves the quality of the obtained data with low precision, while the frame rate decreases. Adjusting the refresh rate parameters can be used to digitize moving objects. A lower number of images reduces the accuracy of the data when digitizing moving objects. Under specific conditions and from a certain speed, the obtained data may be blurred and inaccurate [13].

Selecting the exposure time setting from 62.5 to 1000 μs restricts the amount of reflected light entering the ToF sensor. Exposure time is used when oversaturating an image and reducing shiny areas that result in inaccurate points in front of or behind the actual subject. Exposure time adjustment is combined with conversion gain signal gain tuning. The lower exposure and the higher gain are used in case of reduced accuracy due to multiple light reflections from multiple surfaces called multipath. The conversion gain setting with the high and low option is used when empty areas are created by low-reflective surfaces, while the reflected signal is weak and needs to be amplified. Higher exposure time and conversion gain values are used to digitize objects at greater distances or with low reflectivity. Lower values are used to digitize close and shiny objects [13].

The spatial filtering setting reduces inaccuracies by adjusting and averaging the differences in the values of the coordinates of neighboring points, thus smoothing the surface. This parameter is often used to improve edges in complicated shape objects where a high level of detail is not required. The confidence threshold setting is used especially in cases where it is not possible to eliminate quality deficiencies in the acquired data using exposure tuning and conversion gain caused by reflections and low light reflectivity. In some cases, the light reflected from the surface has a low intensity, therefore the reliability of this signal is limited. The confidence threshold setting allows you to filter reflections from surrounding objects and flying distant pixels [13].

Process of Digitization of Objects The preparation of the digitization can be divided into two parts: hardware, software, and scene preparation. The complexity of preparing a scene depends on the application. The complexity can be understood according to the chosen communication interface and the method of powering the Helios2 device [14]. For the purposes of the digitization, the method of power supply and data transmission via GigE and M12 X-coded interface connected with PoE + IEEE 802.3 at device Tenda PoE30G-AT and laptop for working with GUI and camera data is chosen. The Helios2 device was attached to a tripod at a height of 800 mm from the floor, to which it was attached using an M6 screw and custom holder made by FDM technology. For safety reasons, the tripod was secured against tipping over by attaching it to the floor. After connection, the Helios2 device was allowed to warm up to its own operating temperature for 20 minutes. In order to get closer to the real industrial conditions, the laboratory was isolated from outside light and illuminated by an internal light source from light bulbs. The room temperature was constant at 20 ° C during the experiment. Simple installation of the ArenaView software available on the manufacturer's website was required. As part of the scene preparation, it was necessary to consider the number and material of the surface of the selected digitized objects and the environmental conditions within the digitization of the scene. At the beginning of the experiment, white chapter was chosen as the basis for the digitization objects in order to demonstrate the function of the device against the white color of the surroundings. The camera made an angle

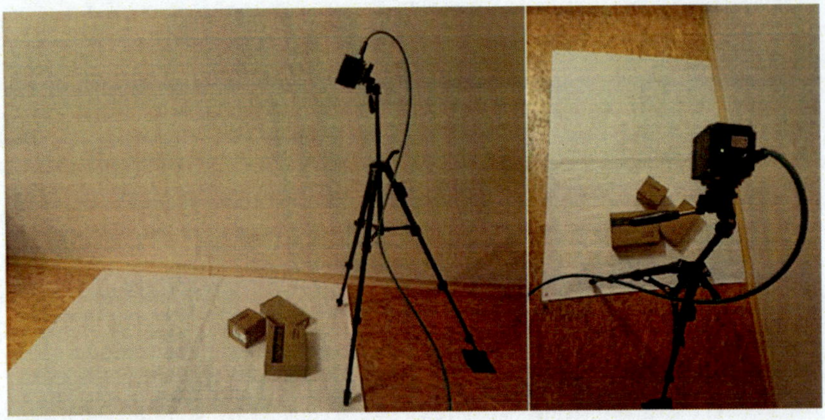

Fig. 2 Scene preparation for digitization

of 30° to the tripod during the experiment, and the digitized objects were placed in the camera's field of view. Within the selection of digitized objects, objects of cardboard were selected as one of the most common materials used in bin picking, palletizing, or robot and computer vision (Fig. 2).

The following factory default settings have been selected for the initial settings of the Helios2: Acquisition Frame Rate 30 MHz, Exposure Time Selector 1000 μs, Pixel Format Coord3D_ABCY16, Operating Mode Distance 5 m, Image Accumulation 1, Conversion Gain High, Spatial Filter False, Flying Pixel Removal False. The Depth Range & Color Settings range of 700–1800 mm was selected for 3D data display [15, 16].

Natural color cardboard material can be considered a suitable material for ToF digitization, because it is made of brown color and is balanced in the ratio of light absorption and reflection. Within the display of the digitization in the GUI ArenaView, it is possible to easily register the low quality of the process caused by the selected parameters. First, it is possible to register low accuracy due to image oversaturation in the area of light reflection from the white ambient material due to the white color of the material, the low distance to the camera, and the low angle to the environment and the camera. At the same time, it is possible to observe distant points above the areas formed by the black material of the box and several flying pixels distant from the digitized surface created by random reflection of light against the surroundings. Several parameters have been adjusted to compensate for image oversaturation. Enabling confidence threshold would completely remove these areas, from which it can be concluded that the signal strength is low in these areas. Changing gain to low increased the quality of the data in these areas, creating small areas of low precision at the edge of the object caused by multiple reflections of multipath light from the wall and substrate. Changing the exposure time to 250 μs increased the data quality of both of these areas, while reducing the accuracy of the data to a minimal extent. This reduction in precision was corrected by increasing

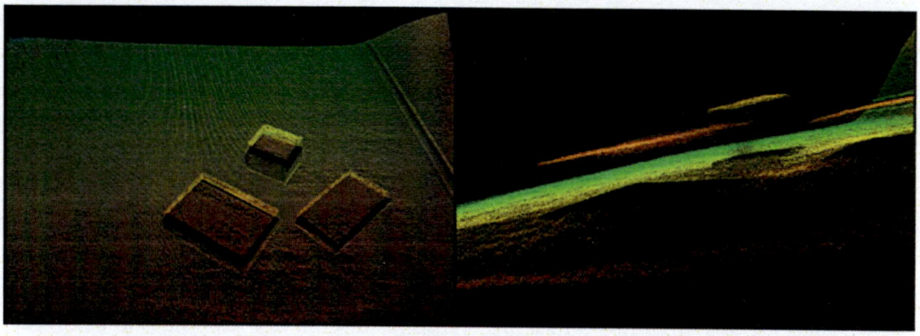

Fig. 3 Digitized surface of objects from natural cardboard before parameter adjustment

Fig. 4 Digitized surface of objects from natural cardboard after parameter adjustment

image accumulation to 16 because the objects are static and the reduction in frame rate at the cost of higher data quality is fine. Distant points in areas of black material are compensated by the spatial filter option, which has increased precision in these areas. Several distant points outside the scene were compensated by the flying pixel removal option. After adjusting the parameters, the point cloud was exported to the ply format so that it could be used for subsequent work in the GOM Inspect system (Figs. 3 and 4).

3 Polygonization of Digitized Surface Using GOM Inspect

3.1 GOM Inspect

GOM is a German company producing measuring technology since 1990. The company develops, produces, and distributes software, measuring devices, and systems based on 3D coordinate measuring technology for various industries. GOM develops software for analyzing 3D data obtained from GOM systems, 3D scanners, laser scanners, and others. GOM Inspect includes import of measurement data, import of basic CAD formats, operation of measuring systems, alignment and use of local coordinate systems, comparisons, measurement of geometric tolerances, inspection of curves, analysis of surface defects, report export, parametric inspection, import native CAD formats of known systems, use of templates for projects, use of digital assemblies of components, scripting, virtual measuring space, and others. GOM Inspect is based on a parametric approach providing the ability to document steps and results for reports. This software works on the principle of creating 3D polygon meshes from a point cloud of digital-isolated objects scanned by measuring devices. Polygon mesh describes the surface of real objects and is used for visualization, simulation, reconstruction, and comparison of digitized data with nominal CAD models. Polygon mesh can be smoothed and recalculated, holes filled in, all with a view of deviations in real time. A very useful comparison used in industry is the point comparison of polygon network deviations against nominal CAD data [17, 18].

3.2 Polygonization Process

Polygonization is a process performed within the software capabilities of CAD systems, during which wireframe surface 3D models are created based on the systematic connection of coordinate points from a point cloud by polygons, i.e., straight lines in space bounded at both ends by coordinate points. Such a polygon meshing is based on specified conditions, which can usually be set at the beginning of the process. These input conditions determine in particular the range of distances of the connected points, the angle between them, the transient boundary conditions, and the noise filtering. If the points in space are too close to each other, or the point is captured outside the specified range of distances, these points are filtered and are not used in polygonization. If there is a large distance between the points, an empty area within the surface may be created. The surface model is most often polygonized into a triangular network, whose advantage is the possibility of describing even more complex shaped surfaces and good compatibility with CA systems, or with slicer software for additive manufacturing [19, 20].

The output within the polygonization process is most often saved in the Standard Triangulation Language format (STL). A higher number of triangles within a polygon mesh allows for more faithful copying of the surface of a real physical model at the cost of a larger file size and the need for higher computational power when working with the model. The appropriate number of triangles is selected automatically within the software solution, or it can be entered manually [21].

After creating a polygon model, it is possible to perform further processing to increase its quality characteristics and usability. Various CA systems provide their mesh processing tools. Common tools and methods for processing polygon models include removing the excess surrounding area of the digitized model, removing edges at an angle from the desired surface, filling empty areas of the model

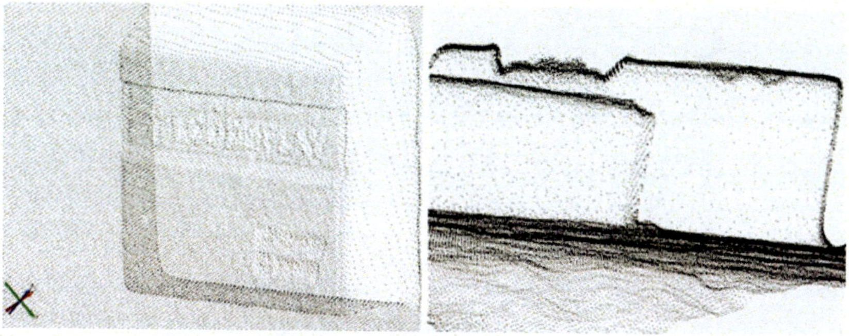

Fig. 5 Local reduction of accuracy of acquired data in black areas

with surface approximation at the edge of this area, smoothing out meshes and recalculating polygon density [22, 23].

Polygonization Process Using GOM Inspect The first step is to import the digitized data with the Helios2 device. Within the display of the point cloud in the graphical interface environment of the GOM Inspect system, it was possible to recognize the shortcomings of the digitization process. The first type of shortcomings was caused by the black surface printed on the objects, as the parameters of the camera were set to digitize the cardboard, while in these areas there were changes in the characteristics of light reflection relative to the rest of the object. Insufficient light reflection caused by light absorption by a matte black surface caused a local decrease in the accuracy of the obtained data in these areas. The acquired distance of points in these areas was moved further away from the camera (Fig. 5) [24].

Another disadvantage of digitization is the multipath light reflection effect. It happens when emitted light reflects multiple times before being received by the ToF sensor, as opposed to reflecting once. Multipath often happens because of concave angles between object walls and the bottom pad, increasing travel distance of the light, thus increasing flight time. The optical paths of the ToF emitted light combine and interfere with one another (Fig. 6).

Unfiltered flying pixels were removed before starting the data polygonization process. The first step in the polygonization process was to compute normals. This tool performs an automatic calculation of the orientation of normal vectors, so it will then be possible to determine the orientation of the area model obtained by polygonization. This is followed by the polygonization of the data. Polygonise Point Cloud feature in GOM Inspect can be modified by these three parameters – Min. distance of used points, Max. noise, and Max. length of new edges.

Using Min. distance of used points it is possible to define the minimum distance between the points to be polygonized. Larger value of this parameter results as a rougher polygon mesh. Max. noise parameter filters distant points. Using the Max. length of new edges it is possible to specify the maximum length of new polygons. If the distance between the selected points is larger than the chosen edge length,

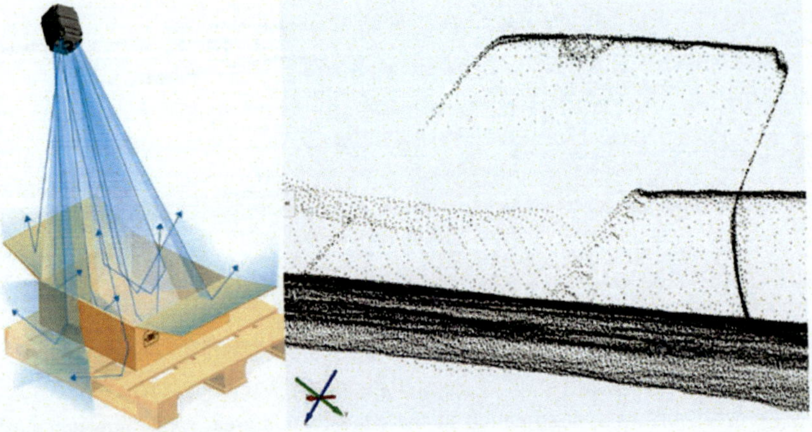

Fig. 6 Example of the multipath effect (left) and its presence in digitized data

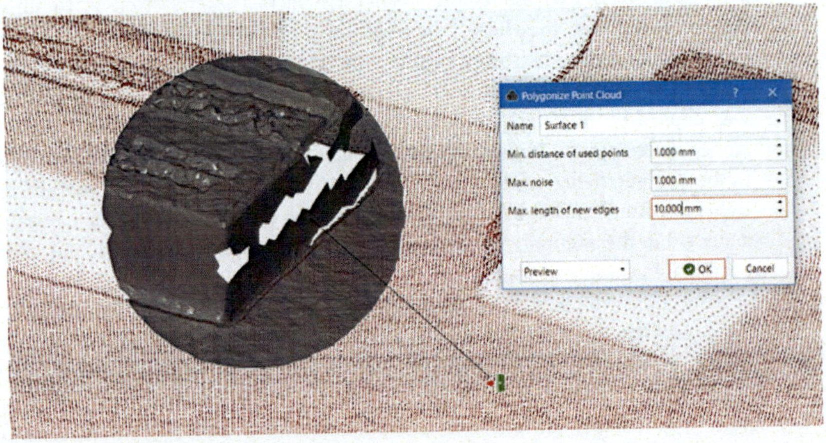

Fig. 7 Demonstration of point cloud polygonization with initial settings

holes result in the mesh. A preview of the polygonization according to the current settings is displayed in the selected circular area (Fig. 7).

The initial parameter values needed to be adjusted. Within the model, empty spaces were created, especially in the areas of stein objects at an angle, because in these areas there is a high distance between neighboring points due to the given camera resolution. Empty spaces could be removed by increasing the value of the Max. length of new edges parameter to a distance of 50 mm. Subsequently, it was necessary to increase the value of the parameter Max. noise, because in the matte black areas the noise was higher than the original 1 mm. This parameter has been adjusted to 3 mm. Min. distance of used points can be used to smooth the model.

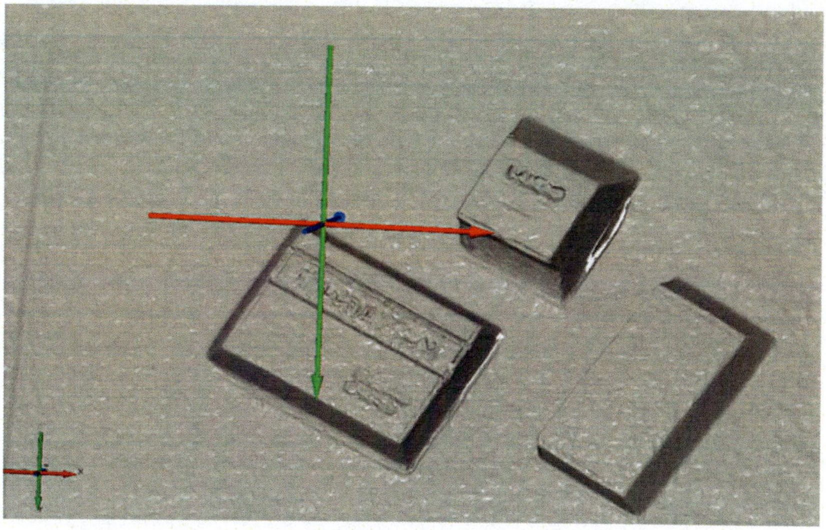

Fig. 8 The resulting model of the polygonization process

The precision of the camera within the digitized scene, specifically Operating Mode Distance 5 m and surface distance from the camera about 1 m, the value was 0.6 mm. Above this distance, it is possible to smooth the surface of the obtained model. Within the balance between accuracy and quality of the polygonized model, the value of the parameter 2 mm was chosen (Fig. 8).

4 Conclusion

This chapter discusses the 3D time-of-flight digitization solution Helios2 along with the ArenaSDK software package, point cloud acquisition, and polygonization within GOM Inspect. The aim of this research was to verify the use and functionality of the camera for the needs of digitization of objects. The first part of the article describes the Helios2 device, its physical and software capabilities, and important features of the ArenaView GUI which affect the quality of the obtained point cloud. The second part describes polygonization process and software GOM Inspect. The preparation of the digitization was divided into hardware, software, and scene preparation. Three boxes made of cardboard with natural color were digitized. Subsequently, the obtained surface was imported into the GOM Inspect software and polygonized. Polygonization parameter values have been adjusted. The result was a polygonal model of the surface. In the process of polygonization, a decrease in the quality of the obtained data occurred caused by the field of view of the camera, black matte material, and multipath effect. The advantage of using time-of-flight technology for digitizing objects is the time needed for data acquisition, which can be obtained from 250 μs. Another advantage is the fact that Helios2 is a plug & play and user-friendly IP68 industrial device. The disadvantages are the low accuracy and precision of the obtained data. The obtained data quality is sufficient for computer vision and robot vision, because with the use of the obtained data it is possible to recognize the height and orientation of objects. Further research is needed for accurate and detailed digitization of objects using time-of-flight technology.

Acknowledgments The authors thank the Ministry of Education of Slovak Republic for supporting this research by the grant VEGA no. 1/0051/20, KEGA no. 038TUKE-4/2021 and project APVV-18-0316.

References

1. Van Nieuwenhove, D., Van der Tempel, W., Grrotjans, R., & Kuijk, M. (2006). Time-of-flight optical ranging sensor based on a current assisted photonic demodulator. In *Proceedings Symposium, IEEE/LEOS Benelux Chapter*. IEEE.
2. Lange, R., & Seitz, P. (2001). Solid-state time-of-flight range camera. *IEEE Journal of Quantum Electronics, 37*(3), 390–397.
3. Geng, J., & Xie, J. (2014). Review of 3-D endoscopic surface imaging techniques. *EEE Sensors Journal, 14*(4), s945–s960.
4. Hansard, M., Lee, S., Choi, O., & Horaud, R. (2013). *Time-of-flight cameras: Principles, methods and applications*. Springer Publishing Company, Incorporated.
5. Li, L. (2014). *Time-of-flight camera – An introduction*. Texas Instruments.
6. Becvar, J. (2009). *Vision, J. B. L. GigE Vision cameras and network performance*. https://cdn.weka-fachmedien.de/whitepaper/files/121_leutron_wp_network-performance.pdf. Accessed 28 April 2022.
7. Helios2. LUCID Vision Labs Inc. (2021). https://thinklucid.com/helios-time-of-flight-tof-camera/. Accessed 20 April 2021.
8. Helios2 Time of Flight (ToF) IP67 3D Camera. LUCID Vision Labs Inc. (2021). https://thinklucid.com/product/helios2-time-of-flight-imx556/. Accessed 20 April 2021.
9. Li, L., Xiang, S., Yang, Y., & Yu, L. (2015). Multi-camera interference cancellation of time-of-flight (TOF) cameras. In *2015 IEEE International Conference on Image Processing (ICIP)* (pp. 556–560). IEEE.
10. Falie, D., & Buzuloiu, V. (2008). Distance errors correction for the Time of Flight (ToF) cameras. In *2008 IEEE International Workshop on Imaging Systems and Techniques* (pp. 123–126). IEEE.
11. Arena SDK Documentation. LUCID Vision Labs Inc. (2021). https://support.thinklucid.com/arena-sdk-documentation/. Accessed 22 April 2022.
12. Helios2 ToF 3D Technical Reference Manual. LUCID Vision Labs Inc. (2021). https://support.thinklucid.com/helios2-hlt003s/. Accessed 20 April 2022.
13. Time of Flight Guidebook. LUCID Vision Labs Inc. (2021). https://thinklucid.com/time-of-flight-camera-guidebook/. Accessed 22 April 2022.
14. Francis, S. L. X., Anavatti, S. G., & Garratt, M. (2011). Reconstructing the geometry of an object using 3D TOF Camera. In *2011 IEEE Workshop On Merging Fields of Computational Intelligence and Sensor Technology* (pp. 13–17). IEEE.
15. Oprisescu, S., Florea, L., & Ovreiu, E. (2013). Detection of thrown objects using ToF cameras. In *2013 IEEE 9th International Conference on Intelligent Computer Communication and Processing (ICCP)* (pp. 83–86). IEEE.

16. Oprisescu, S., Rasche, C., & Su, B. (2012). Automatic static hand gesture recognition using ToF cameras. In *2012 Proceedings of the 20th European Signal Processing Conference (EUSIPCO)* (pp. 2748–2751). IEEE.
17. GOM Inspect. https://www.capture3d.com/3d-metrology-solutions/3d-software/free-gom-inspect. Accessed 22 April 2022.
18. GOM Inspect Pro. https://www.gom.com/en/products/gom-suite/gom-inspect-pro. Accessed 22 April 2022.
19. Yoo, J.-D., Cho, J.-H., Jang, I.-Y., & Lee, K. H. (2011). 3D content generation using a TOF camera and HDR light information. In *2011 3DTV Conference: The True Vision – Capture, Transmission and Display of 3D Video (3DTV-CON)* (pp. 1–4). IEEE.
20. Li, H., Zou, J., Li, Y., Xu, Y., & Xiao, Z. (2020). TOF camera array for package volume measurement. In *2020 7th International Conference on Information Science and Control Engineering (ICISCE)* (pp. 2260–2264). IEEE.
21. Wang, W. (2010). *Reverse engineering: Technology of reinvention* (360p). CRC Press. ISBN 978-1439806302.
22. Lee, C., Song, H., Choi, B., & Ho, Y.-S. (2011). 3D scene capturing using stereoscopic cameras and a time-of-flight camera. *IEEE Transactions on Consumer Electronics, 57*(3), 1370–1376.
23. Pal, B., Khaiyum, S., & Kumaraswamy, Y. S. (2017). 3D point cloud generation from 2D depth camera images using successive triangulation. In *2017 International Conference on Innovative Mechanisms for Industry Applications (ICIMIA)* (pp. 129–133). IEEE.
24. Vodilka, A. (2020). *Analýza voľného odpruženia hliníkových výliskov a rekonštrukcia výlisku po lisovacej operácií.* Prešov. Diploma thesis, Technická univerzita v Košiciach, Fakulta výrobných technológií so sídlom v Prešove.

Application of Smart Construction Site in Construction Practice

Jana Smetanková, Katarína Krajníková, Rastislav Ručinský, Peter Mésároš, and Annamária Behúnová

1 Introduction

Construction is an important sector of the economy. It is an economic sector engaged in the construction and management of construction and commercial activities. The basic tasks of construction include the production, sale of building materials, organization, management, and preparation of construction activities.

In the European Union, construction provides about 18 million direct jobs, which is about 9% of the European Union's gross domestic product. Construction creates new places of work, supports economic growth, and provides solutions to social, climate, and energy challenges. The main challenges for construction include:

J. Smetanková (✉)
National Infrastructure for Technology Transfer Support in Slovakia II – NITT SK II, Bratislava, Slovakia

Technical University of Kosice, Faculty of Civil Engineering, Expert's Institute in Construction, Kosice, Slovakia
e-mail: jana.smetankova@tuke.sk

K. Krajníková · P. Mésároš
Technical University of Kosice, Faculty of Civil Engineering, Institute of Construction Technology, Economics and Management, Kosice, Slovakia
e-mail: katarina.krajnikova@tuke.sk; peter.mesaros@tuke.sk

R. Ručinský
Technical University of Kosice, Faculty of Civil Engineering, Expert's Institute in Construction, Kosice, Slovakia
e-mail: rastislav.rucinsky@tuke.sk

A. Behúnová
Technical University of Kosice, Faculty of Mining, Ecology, Process Control and Geotechnologies, Institute of Earth Resources, Kosice, Slovakia
e-mail: annamaria.behunova@tuke.sk

L. Knapčíková, D. Peraković (eds.), *7th EAI International Conference on Management of Manufacturing Systems*, EAI/Springer Innovations in Communication and Computing, https://doi.org/10.1007/978-3-031-22719-6_4

- Stimulating demand – the aim is to improve activities in existing buildings through renovations, thus directly stimulating potential and demand.
- Training support – the aim is to improve specialized training, to make the sector more attractive, especially for workers, technical colleges, and universities.
- Support for the implementation of innovations – the aim is to streamline the process of adopting new technologies and materials.
- Increasing energy efficiency and eliminating climate impacts (climate change) – the aim is to build buildings that reduce energy burdens (buildings have the largest share of total final energy consumption in the EU (40%) and produce around 35% of all greenhouse gas emissions) [1].

Developments in all industries and societies are constantly advancing. The construction industry has a long history of innovations that support progress in construction implementation and management. After many years of growth and increasing development and implementation of innovations, there has been a difficult year caused by a pandemic. The crisis has affected the development and implementation of construction projects. Individual companies tried to protect their employees and re-create jobs. The industry has responded to the difficult period with an increased focus on innovation, bringing progress in the development of automation and technology, which persist in construction and become an important part of the individual processes [2].

Applying innovation in construction is not easy, despite the direct impact on industry development and economic growth. Each construction project needs to be solved individually, i.e., to adapt individual processes and resources. Construction processes and works are located in different places and involve the constant movement of machines and personnel. A company that has an implemented innovation management process can gain several significant benefits, such as:

- Improving the organization
- Improving competitiveness in the medium and long term
- Improving the configuration of management processes and overall meeting
- Streamlining the use of knowledge – knowledge management
- Improving client satisfaction [3]

The most significant innovative technologies, which are gradually implemented into individual processes in the planning, implementation, and use of buildings, include augmented, mixed or virtual reality, building information modeling, construction exoskeletons, construction robots, drones, artificial intelligence and machine learning, modular constructions, 3D printing, and many other technologies.

2 Smart Construction Site in Construction Practice

A construction site is a space that is intended for the performance of construction work on a construction site, storage of construction products and transport, and other equipment necessary for the construction and for the location of construction site equipment. Construction site equipment is the tools needed to carry out construction work, to carry out alterations to construction sites or for maintenance work, and serves an operational purpose, a production purpose, a storage purpose, and a social purpose. The construction site must meet certain requirements and requirements, such as

- It must be secured against the entry of foreign persons into places where life or health may be endangered.
- Must be marked as a construction site with the necessary data on the construction and the participants in the construction.
- Have set up entry and exit from the local or special-purpose road for access to construction products, for the removal of land and construction waste, and for the entry of medical and fire protection vehicles to be cleaned.
- Enable the safe storage of construction products and construction mechanisms and the location of site equipment.
- Enable the safe movement of persons carrying out construction work.
- Have waste disposed of or disposed of.
- Have the equipment necessary to carry out the work and the housing which carries out that work.
- The construction site must be set up and operated in such a way as to ensure the protection of human health on and around the construction site, as well as the protection of the environment in accordance with special regulations [4].

For these requirements and requirements, the construction sector is constantly developing innovative ways and tools that would streamline individual construction processes and procedures. Based on the above, the tools and principles of a smart construction site are increasingly being applied in construction practice.

The intelligent construction site consists of interconnected systems, sensors, displays, and computer technology that operate on the basis of advanced applications. The intelligent construction site model is designed to implement real-time ICT connectivity, enable mutual recognition, and streamline communication between workers, machinery, the construction environment, and material resources [5].

The implementation of smart construction tools and procedures brings a number of benefits that improve various aspects, such as time and financial planning, increases the productivity and safety of construction projects and employees, and the like. The biggest advantages of applying a smart construction site in construction practice include the following aspects:

- Optimizing the use of construction tools and machines by monitoring their location
- Improving the quality of building materials by monitoring their implementation, for example, by monitoring concrete hardening processes or detecting changes in temperature, humidity, and the like
- Improving transparency (analysis of construction development – construction progress reports)

- Increasing safety on the spot, for example, by detecting when workers are exposed to high noise, poor air quality, and so on
- Reducing cyber attacks and eliminating data leakage [6]

The biggest barriers to the implementation of smart construction tools and procedures include:

- High input costs (poorly supported development of innovations in the construction industry)
- The need for professional training (increased travel costs, knowledge of a foreign language, time-consuming, etc.)
- Ignorance and unwillingness of the public to implement innovations in their processes [7]

Different types of intelligent tools are known that can be implemented within the aspects of a smart construction site. In general, they can be divided on the basis of the following criteria:

- Tools to ensure safety on the construction site
- Tools to ensure protection of materials and property
- Construction robotics tools
- Tools for managing and supporting the application of intelligent elements on the construction site – construction applications.

2.1 Tools to Ensure Safety on the Construction Site

Within the planning and management of activities on the construction site, a number of aspects, inputs, and parameters enter into individual processes. The principle of monitoring these changes varies within a traditional and intelligent construction site, but the common point and priority is always to monitor the safety of employees and operations. Workers on a construction site are often exposed to many adverse aspects that affect their attention, making them distracted and this can lead to reduced concentration and carelessness. The work performed within the implementation of construction works is characterized by a considerable force load and affects the psychological well-being and concentration of employees. To eliminate the adverse effects on these aspects, many intelligent aids are gradually being implemented that support and increase the health and safety of all employees. These innovative tools include, for example, Smart Cap (SC for short).

Smart Cap is a device for monitoring employees, their location, activities, and security. Through SC, it is possible to monitor the potential risks and energy status of a person in real time in a real place. Active connection with the I-Cloud platform enables efficiency in the structure of changes and optimization of work productivity. A separate headgear serves as a mandatory protective part of the employee's clothing on the construction site, which is complemented by a LIFE BRAND bracelet. It works on the principle of monitoring brain waves, the so-called

EEG (or electroencephalogram) and serves to inform human alertness. Operators can monitor employee vigilance through an application that allows you to visualize and audibly connect an employee to a manager via a smartphone or computer. The application itself works on the principle of Bluetooth and informs the management about the current use of manpower on the construction site [8, 9].

Excessive strain and fatigue in the workplace are common factors in incapacity for work and labor shortages in the workplace. Exoskeletons (abbreviated EXO) are wearable electromechanical devices with motorized joints, which are placed on the user's body, act as an amplifier to correct posture, weight dissipation, and lift support, thus contributing to the efficiency of the work. They are made of steel, which is coated with reinforcing material for more comfortable wearing. The device is controlled manually or using electricity, or a combination of both. The aid can also be supplemented with various types of sensors and hydraulic systems. There are many types of EXOs to use, such as the arm and shoulder exoskeleton (a cushioned vest that is designed to support the hands when working overhead), the spine support exoskeleton (coordinates proper posture when performing work), and useful exoskeletons for long-term work in standing position (providing standing support) and a mechanical exo-skeleton arm (used to secure tools and support when lifting heavy machinery using scaffolding and a work platform) [10–12].

Another tool of a smart construction site is mixed reality glasses. The glasses place 3D objects called holograms into the real world and bring a new level of productivity and innovation. The glasses increase productivity in real time and alert and analyze errors more quickly. The use of machinery and space is minimized, increasing productivity and return on investment. At the construction site, the goggles are used primarily for checking the construction, adhering to the time schedule, and checking the completion of individual structures and works [13].

2.2 *Tools to Ensure to Protection of Materials and Property*

The company's assets are an essential input for carrying out construction work. In order to protect property from theft, appropriate measures must be taken, from the most common ones, such as site fencing and material storage, to those specifically such as GPS locators that can be placed on any machine, material, or device.

Increasingly, camera systems are being used to ensure the protection of materials and equipment on the construction site. The construction site is an ideal space for thieves, due to the fact that there are a number of materials, machines, and equipment that have a high financial value. There are a number of quality safety systems for monitoring the construction site, the most well-known being the comprehensive monitoring system of the Videofied construction site. Videofied is a tool that can be operated without the need for electricity and internet connection, uses its own data connection based on mobile signal, triggers recording/recording of movements on the construction site via infrared motion detection, recognizes people and objects (Smart PIR technology) and via a mobile application based on the Cloud system, it sends SMS messages with immediate notification, thanks to which the construction site is under constant supervision [14].

Another tool for monitoring site activities is OSI technology, which continuously and efficiently records all site activities. In the event of an unauthorized intrusion, the technology immediately detects the intruder and sends an alert to the security guard's facility or directly to the police. The device also declares high resistance to extreme conditions, such as explosion, records weather and air quality (high-quality data collection), can be remotely set, panned, tilted, zoomed in and out, and is used for coordination and management over long distances (online via a web browser) [15].

In addition to these camera systems, we also know the thermostatic terrain mapping FLIR, which is used mainly in aviation, but is increasingly used in the construction segment and the Vekarda system, which stores large amounts of data in a large format storage, allows live video and recorded records, provides accurate analysis and identification of employees, objects, and vehicles.

The latest technology used to monitor and record site activities are diamond systems, which provide advanced site monitoring. The system is a simplified platform for processing and securing all sites from a mobile device or web browser. The systems make it possible to create video, record deliveries of materials, provide alerts on the arrival and departure of vehicles, mark the location of materials in equipment and applications, and prevent accidents, loss of materials and equipment during operation [16–18].

In addition to camera systems, GPS locators are also used on the construction site, which are portable devices that allow monitoring of people, animals, vehicles, or objects. They work on the basis of a SIM card in a device that is connected to an application (abbreviated APP) in a phone or software in a computer. Such APPs and software are offered as standard by the manufacturers of such devices, which offer full compactness. Attaching the GPS locator is possible mechanically or simply with a magnet. They can withstand high and low temperatures from −20 to +70 ° C, are water resistant, and have a low wear rate [19].

Classic paper attendance is now increasingly being replaced by intelligent systems that offer a number of functions, designs, and accessories. The systems were created to facilitate the management of the company and the fulfillment of obligations. Attendance solutions can be monitored in mobile APPs resp. on web-accessible sites allowing you to control the employee's work easily and quickly from anywhere. By applying these types of attendance systems on the construction site, we can greatly simplify the work of a manager who completes attendance. In the form of automatic data storage and recording of working employees present on the construction site, we have an effective overview of employees' hours worked, thus eliminating late arrivals and early departures.

Attendance solutions can be divided into three basic categories:

- Hardware attendance systems – operate on a system of capturing a person's face with a camera or by capturing personal physical objects, i.e., by placing a finger or a card, chip, or other devices.

- Cloud attendance systems – they are used mainly in certain types of professions in construction, where it is necessary not only to physically perform work at a designated workplace or also to perform work externally, outside the workplace. The employer needs a full overview of the performance of work in the form of employee records, which is also used by various attendance monitoring solutions, for example by connecting monitoring of company equipment – mobile phones, tablets, computers, etc.
- Combined attendance systems – the most commonly used systems that record on-site attendance. It is a connection of a hardware device with an application that brings simple and convenient attendance editing [20].

2.3 Construction Robotics Tools

Robotics is an ideal way to improve productivity and cost efficiency. In construction, robotics has built a reputation as a slow-growing and expensive construction solution. The most innovative modern form of using small robotics is the introduction of the use of drones or monitoring equipment such as the robotic dog Spot for full monitoring and ensuring the safety and fluidity of work on the construction site. The most common form of automation in construction in the field of small construction robots is the introduction of the use of drones, which plays a key role in safety and efficiency. Drones are important for planning and monitoring the effectiveness of construction, the development of the progress of work, which are an integral part of the detection of undesirable problems, and obstacles that can disrupt the construction schedule [21].

In addition to small construction robots, large painting robots and masonry robots are also used. The masonry robot lifts the brick, automatically applies the grout, then places the brick in place with great accuracy in millimeters, balances it as needed, and places it so that it exactly meets the required standards of the given technological regulation of masonry material [22].

Reinforcement design requires precise reinforcement of the reinforcement with wire, which increases the strength of the structure. For this purpose, robotic reinforcement connectors have been developed that work and move on a rail system and can create up to 1100 connectors per hour [23].

Autonomous heavy machinery is automotive machinery. Most of these machines work with a system of cameras and sensors. The need for machinery in the construction industry is undoubtedly an integral part of the need to carry out work on the construction site. Construction of all levels of difficulty requires optimization of demands, constraints, delays, and safety. Among the usable machines are paria excavators, bulldozers, loaders, manipulators, and motor graders, which operate on different propulsion principles and in different sizes, taking into account the size of the construction site and the volume of work.

2.4 *Tools for Managing and Supporting the Application of Intelligent Elements on the Construction Site: Construction Applications*

Artificial intelligence in the construction industry is designed to create construction improvements and speed up correct and effective problem-solving decisions. Recently, there has been great progress and development of systems in the field of construction with applications mainly in the field of project evaluation, diagnostics, safety, human resources, architecture and building optimization, project management technology, road and bridge health identification, and other specialized areas. We distinguish several types of APP that are useful for the construction industry:

- APP for management and control (Cat MineStar Edge, Measurement, Solocator, Contractor Tool Tracking, Madcat, and others) [24–26]
- Narrowly profiled project applications that fulfill one function, but in the most widespread format (AutoCAD, GNACAD, DWG FastView – CAD) [27–31]
- Digital assistants and applications for conversions and calculators in construction (Plaradar, Fieldrive, Procore, Bsgapp) [32–34]
- Managerial and team APP helping to monitor and optimize human resources (Buildo, Buildary.Online, Asana, Wrike, and others) [35–38]

3 Methodology

3.1 *Research Aim*

The research was focused on examining the current state of use of smart construction site tools in Slovak construction practice. The aim was to analyze the approaches and use of smart construction site tools in the processes of construction work on the construction site. The research analyzes general information related to the use of smart tools, specifically focusing on the level of knowledge of the concept of smart construction site, the use of tools to ensure the construction site, health and safety of employees, the use of tools and their degree of implementation within the construction site.

3.2 *Data Collection and Research Sample*

Construction companies operating in Slovakia were contacted through an online questionnaire survey. The aim of the research was to address the general professional public, i.e., companies operating in the public and private sectors. Priority was given to a questionnaire survey focused primarily on construction companies, i.e., suppliers (main contractors, subcontractors) of construction works. As part of the survey, 146 companies were contacted by e-mail. Of the total number of companies and organizations contacted, the return rate of the questionnaire was approximately at the level of 33.58%. The reason for the relatively low return was mainly neo-willingness (high number of parallel online surveys), high workload of companies, and lower awareness.

3.3 *Research Step and Methodology*

At the beginning of the questionnaire, the respondent was acquainted with the purpose of the questionnaire and the approximate time frame for its completion. The questionnaire itself consisted of 34 questions and was divided into 4 parts:

- Section 1 – information on the respondents – type of construction participant, type of organization, sector of operation, size and nature of the company, and length of operation on the market.
- Section 2 – degree of knowledge of the concept of smart construction site – concept (definition of the concept), use of tools.
- Section 3 – use of smart construction site tools – tools to ensure property protection, health and safety on the construction site, use of construction robots and applications for the management and application of intelligent construction site tools.
- Section 4 – implementation of smart construction site tools in practice – use, degree of implementation, defining the main obstacles and benefits of applying the principles of smart construction site.

4 Results: Current State of Implementation of Smart Construction Site Tools

4.1 *Information About the Respondents*

The aim of this section was to obtain information about the respondents. The research was conducted in the period April–May 2022. In terms of information about the company, the following data were obtained:

- Type and size of organization
- Sector of operation – market (regional division) Duration of operation on the market
- Focus and character of the company

In terms of type of organization, limited liability companies were the most represented (37.5%), joint-stock companies (37.5%), followed by self-employed persons (self-employed persons), specifically 18.8% of respondents and public companies (6.3%). In terms of the size of construction companies in which they carry out their activities (classification by number of employees – classification according to the recommendation of the European Commission 2033/361/EC), micro-enterprises were most numerous (0–9 employees) – level 31.3%, followed by small enterprises (10–49 employees) and large companies (500 and more employees) with a level of representation of 25%. The companies with the least represented group were defined as medium-sized enterprises (50–249 employees), specifically 18.8% of the respondents represented this. In terms of operational and geographical division, the companies operating in the Košice Region (75%), the Prešov Region (50%), and the Bratislava Region (18.8%) were the most represented. The companies operating in the Žilina, Nitra, and Banská Bystrica regions were at the level of 12.5% of the respondents. The companies operating in the Trnava and Trenčín regions were the least represented, concretely 6.3% from the approached companies. In terms of focus or type of construction projects on which the company focuses, companies representing or specializing san and building construction (73.3%) were represented, and the rest of the respondents focused mainly on the implementation of civil engineering.

4.2 *Degree of Knowledge of the Concept of Smart Construction Site*

After obtaining basic information about the respondents, analyses of the level of knowledge and use of smart construction site tools followed. Respondents answered the questions:

- Have you come across the term smart construction site?
- What do you mean by smart construction site?
- Do you know or have heard of some of these smart construction site tools?
- Do you use any of the selected smart tools in your company?
- How long have you been using smart tools/construction site solutions in your company?

56.3% of the respondents had already encountered the term smart construction site as part of their activities, and under this term they mainly imagined the use of applications and software to make work more efficient (81.3%). Detailed results for the question "What do you mean by smart construction site?" are displayed in Fig. 1.

In terms of knowledge of selected intelligence tools, respondents knew mainly drones – unmanned aircraft, which are remotely controlled and allow scanning and recording of work on the construction site (68.8%), hololens, which project virtual reality into the real environment (50%), smart cap, which serves as a device for monitoring employees, their position, activity, and safety (25%) and the EVO

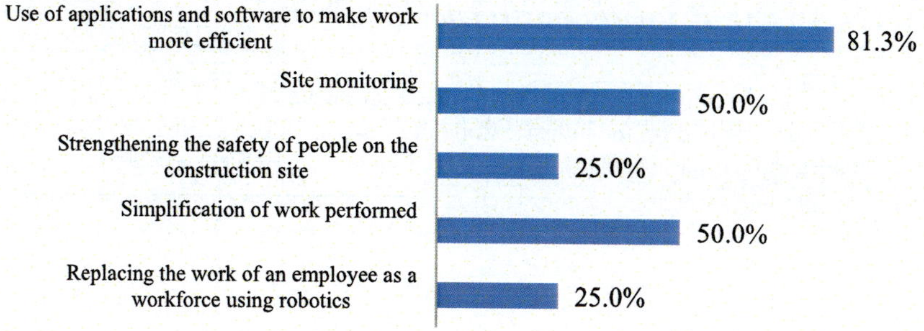

Fig. 1 Response results: "What do you mean by smart construction site?"

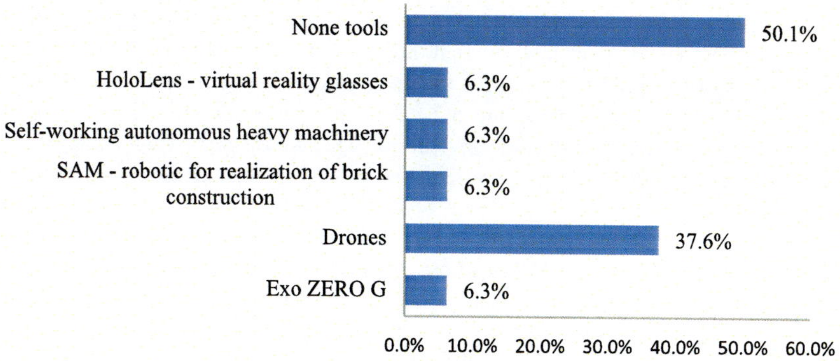

Fig. 2 Do you use any of the selected smart tools in your company?

exoskeleton – a sprung vest, designed to support the hands when working overhead (25%). Respondents also knew the technologies of self-working autonomous heavy machinery (18.8%), TyBOT technology – robotic equipment for binding reinforcement (12.5%) and at the level of 6.3% of respondents were SAM tools – robotic equipment that can use sensors to make a brick construction and Picto Bot technology, i.e., a robotic painter.

From the point of view of active use of tools in construction, the most used are unmanned aircraft, robotic equipment for creating masonry structures, EXO ZERO G skeleton (mechanical arm for lifting and working with heavy tools and machines), and hololens glasses. Detailed results are shown in Fig. 2. Active users of smart tools use them in their companies for about 1–3 years (42.9%), the respective 6 months to 1 year (7.1%), and less than 6 months (14.3%).

4.3 Use of Smart Construction Site Tools

The third section of the survey focused on occupational safety and protection of property placed on the construction site. As part of the survey, we focused on two subsections, namely the level of use of tools ensuring the protection of property, health and safety on the construction site, and the rate of use of applications for the management and application of intelligent tools on the construction.

In the first subsection, it was examined whether the addressed respondents feel tired and exhausted in their performance. Fatigue and feelings of exhaustion significantly affect employee performance, which can lead to accidents and reduced safety. As many as 43.8% stated that they rarely, or not often, feel tired and exhausted (31.3%). About 18.8% feel tired often, but only 6.3% say they don't feel tired at all. Interesting results were achieved in the use of protective equipment, where some respondents saw that they use them not often (6.3%) or rarely (6.3%). This fact is quite surprising, as the use of protective equipment on the construction site is legally enshrined and it is the employee's responsibility to comply with applicable health and safety regulations. Despite the significant benefits of using a smart cap, only 44% of respondents would consider using it in their activities. In the field of site protection and monitoring, up to 62.5% of respondents do not use any of the smart solutions, 31.3% of respondents use the Videofied camera system, and 6.3% use the VERKADA camera system, which automatically identifies people, vehicles, materials, and machines. The survey showed that the use of localization technologies are used only rarely in construction practice. As many as 64% of respondents stated that they do not use them at all and only about 32% use GPS tools and 4% of respondents use the RiFID system.

The second subsection examined the use of applications to manage and deploy intelligent tools on the construction site. The most used are applications supporting managerial and teamwork (31.3%). These were mainly the Planradar applications (21.4%) and the B2W Mobile Construction App (7.1%). This was followed by conversion-enabled applications, so-called calculators (25%), where MADCAD (20%), MILWAUKEEE One-key applications (13.3%), DEWALT Tool Connect, Solocator, MR Builder, CAT MineStar Edge, and Contractor are used. Tooltracking (all at 6.7%). In terms of project applications, 25% of the respondents actively use them, especially the Au-toCad application (75%) and DWG FastView – CAD ViewerEditor (18.8%). The last type of applications used are digital assistants and digital planners, where 12.5% of the addressed users are active users of these applications. It was mainly the use of digital construction diaries. As many as 87.5% of respondents who do not use any digital assistants and digital planners would consider implementing a digital construction diary. The most common way of communication between work teams is verbal communication and communication via mobile phones (81.3%), followed by communication via email (56.3%), SMS messages (25%), and via social networks and tools (Viber, Messenger), and the like (25%).

Table 1 Advantages and barriers to implementation of smart construction site tools

Advantages to implementation of smart construction site tools	%	Barriers to implementation of smart construction site tools	%
Increasing the quality and safety of work	62.5	High costs with the procurement of product intelligent tools	93.8
A new way of organizing the workplace and external relations	43.8	The need to train employees to work with new technologies	32.3
Streamlining construction scheduling (schedules)	62.5	Unavailability of applications and software in the Slovak language	32.3
Introduction of new methods of work organization using applications/artificial intelligence	50.0	Insufficient information about the maturity of construction innovations	50.0
New ways of providing services	25.0	Distrust in the efficiency of robots	6.3
Easier solutions to eliminate deficiencies on the construction site	50.0		

4.4 *Implementation of Smart Construction Site Tools in Practice*

The conclusion of the survey examined the possibilities of implementing smart construction tools into the activities of the addressed companies. Up to 44% of respondents are considering the implementation of a smart construction site and its tools. Twenty-eight percent of respondents stated the possibility of possible and up to 28% of respondents stated that they do not plan and do not want to implement smart construction tools into their activities. Among the respondents who are considering the implementation of smart construction site tools into their activities, they saw that they wanted to implement them into their processes in the time horizon of 1–3 years (45.5%), respectively 3–5 years (36.4%) and less than 1 year (18.2%).

In addition to the degree and willingness to implement smart construction site tools, it was examined how the addressed respondents perceived as the biggest obstacles and benefits of smart construction site implementation in their construction implementation processes. Detailed results are displayed in Table 1.

5 Conclusion

The use of smart construction site tools and approaches is a revolutionary way to do work through digital intelligence tools and applications. By implementing the tools, it is possible to increase work efficiency, improve construction scheduling, safety, communication, and work organization on the construction site. There was a problem with the reluctance to participate in the online survey in the data collection process. In the future, the aim will be to carry out similar surveys in a combined form of data collection, i.e., in the form of personal interviews and an online questionnaire. The level of knowledge of the term smart construction site is average in Slovakia and users are familiar with the technologies of unmanned aircraft (drones), goggles for projecting reality into the real environment (hololens), smart cap, and EVO exoskeletons. The most commonly used are unmanned aircraft for recording the activities and movement of employees on the construction site, robotic equipment for the implementation of individual structures, camera systems, digital attendance systems, and autonomous heavy machinery with GPS and RFID localization. The survey showed a willingness to implement individual tools and concepts of smart construction site into individual activities, but also pointed to possible disadvantages and barriers preventing the rapid implementation of these tools and procedures in the processes of preparation, planning, and management of building construction. The aim of the future research will be to monitor the current development of the use of selected smart tools on the construction site and to analyze the newly obtained data with the data that are the results of this survey. At the same time, it will be important to monitor and point out possible obstacles to the implementation of the selected tools and propose solutions that would help to increase the rate of application of the mentioned tools in the processes of planning, management, and control at the construction site.

Acknowledgments The paper presents partial research results of project APVV-17-0549 "Research of knowledge-based and virtual technologies for intelligent designing and realization of building projects with emphasis on economic efficiency and sustainability" and of project KEGA 009TUKE-4/2022 "An interactive tool for designing a safe construction site in an immersive virtual reality." This chapter was created thanks to support under the Operational Program Integrated Infrastructure for the project: National infrastructure for supporting technology transfer in Slovakia II – NITT SK II, co-financed by the European Regional Development Fund.

References

1. European Commission. (2022). *Construction sector*. Retrieved from https://www.bigrentz.com/blog/construction-technology. Accessed 15 June 2022.
2. Stannard, L. (2022). *Construction technology to watch in*. Retrieved from https://www.bigrentz.com/blog/construction-technology. Accessed 10 June 2022.
3. Designing Buildings. *Innovation in construction project*. Retrieved from https://www.designingbuildings.co.uk/wiki/Innovation_in_construction_projects. Accessed 14 June 2022.
4. Lewik. *Construction site (building law)*. Retrieved from https://www.lewik.org/term/10299/stavenisko-stavebny-zakon/. Accessed 5 June 2022.
5. Zhou, H., Wang, H., & Zeng, W. (2018). Smart construction site in mega construction project: A case study on islad tunnelling project of Hong Kong-Zhuhai-Macao Bridge. *Publisher Frontiers of Engineering Management, 5*(1), 78–87. https://doi.org/10.15302/J-FEM-2018075. Retrieved from https://journal.hep.com.cn/fem/EN/10.15302/J-FEM-2018075. Accessed 3 June 2022
6. Technology Cards. (2022). *Smart construction site*. Retrieved from https://www.technologycards.net/the-technologies/smart-construction-site. Accessed 5 June 2022.

7. Liu, H., Song, J., & Wang, G. A. (2021). Scientometric review of smart construction site in construction engineering and management: Analysis and visualization. *Sustainability, 13*(16), 8860. Retrieved from https://www.mdpi.com/2071-1050/13/16/8860. Accessed 2 June 2022
8. Mining3. (2022). *Smart cap*. Retrieved from http://www.mining3.com/wp-content/uploads/2016/07/SmartCap-A4-Factsheet_DIGITAL.pdf. Accessed 28 May 2022.
9. Smart Cap Tech. (2022). *Life smart cap*. Retrieved from http://www.smartcaptech.com/life-smart-cap/. Accessed 28 June 2022.
10. Eksobionics. (2022). *Ekso lifts your construction team*. Retrieved from internet https://eksobionics.com/construction/. Accessed 19 June 2022.
11. Schenker, D. B. (2022). *Exoskeletons: Robotic suits for healthier working conditions*. Retrieved from https://www.dbschenker.com/sk-sk/o-n%C3%A1s/tla%C4%8Dov%C3%A9-centrum/global-stories/exoskelety. Accessed 29 May 2022.
12. Eksobionics. (2022). *Ekso Zero G*. Retrieved from https://eksobionics.com/eksozerog/. Accessed 1 June 2022.
13. Ayes. (2022). *Microsoft hololens*. Retrieved from https://www.ayes.cz/chytre-bryle/microsoft-hololens/. Accessed 1 June 2022.
14. Resideo. (2022). *Residental security*. Retrieved from https://www.resideo.com/gb/en/solutions/security/residential-security/videofied/. Accessed1 June 2022.
15. Safe and Sound. (2022). *Axis security cameras*. Retrieved from https://getsafeandsound.com/axis-security-cameras/. Accessed 1 June 2022.
16. Safe and Sound. (2022). *Flir authorized dealer*. Retrieved from https://getsafeandsound.com/flir-authorized-dealer/. Accessed 1 June 2022.
17. Safe and Sound. (2022). *Rhombus systems security*. Retrieved from https://getsafeandsound.com/rhombus-systems-security/. Accessed 2 June 20222.
18. Safe and Sound. (2022). *Verkada security*. Retrieved from https://getsafeandsound.com/verkada-security/. Accessed 2 June 2022.
19. Alza. (2022). *GPS locators guard everything that is dear to you*. Retrieved from https://www.alza.sk/gps-lokatory-postrazia-deti-psy-ci-auta#ako-vybrat. Accessed 2 June 2022.
20. Voltyla, M. (2021). *Attendance systems*. Retrieved from https://www.podnikajte.sk/technologie/dochadzkove-systemy. Accessed 30 May 2022.
21. Morrison, R. (2021). *The rise of autonomous construction equipment*. Retrieved from https://www.geospatialworld.net/blogs/the-rise-of-autonomousconstruction-equipment/. Accessed 28 May 2022.
22. Madsen, A. J. (2017). *The SAM100: Analyzing labor productivity*. Retrieved from https://digitalcommons.calpoly.edu/cgi/viewcontent.cgi?article=1275&context=cmsp. Accessed 28 May 2022.
23. Advanced Construction Robots. (2021). *Tybot*. Retrieved from https://www.constructionrobots.com/tybot. Accessed 28 May 2022.
24. Autodesk. (2021). *Unlock the power of AutoCAD wherever you go*. Retrieved from https://www.autodesk.com/products/autocad-mobile/overview?term=1YEAR&tab=subscription. Accessed 27 May 2022.
25. Google Play. (2021). *GnaCAD*. Retrieved from https://play.google.com/store/apps/details?id=com.gna.cad&hl=sk&gl=US. Accessed 29 May 2022.
26. DWG FastView. (2021). *DWG FastView for mobile*. Retrieved from https://en.dwgfastview.com/cloud/mobile/. Accessed 29 May 2022.
27. Cat. (2016). *Edge*. Retrieved from https://www.cat.com/en_US/products/new/technology/edge/edge/114700.html. Accessed 10 June 2022.
28. YouTube. (2019). *Clinometer, spirit level, protractor, bubble level for android and iOS*. Retrieved from https://www.youtube.com/watch?v=XpkKryEV7o8. Accessed 1 June 2022.
29. Google Play. (2021). *Solocator – GPS field camera*. Retrieved from https://play.google.com/store/apps/details?id=com.solocator&hl=sk&gl=US. Accessed 3 May 2022.

30. Monday Construction. (2019). *Construction managment softvare that's easy to sed up, and simple to use*. Retrieved from https://monday.com/lp/mb/construction/comparison?utm_medium=cpc&marketing_source=adwordssearch&marketing_campaign=row-sconstruction_management-e-deskmonday&aw_keyword=contractor&aw_match_type=p&cluster=construction&subcluster=&ati=&gclid=CjwKCAjwu_mSBhAYEiwA5BBmfzhp4-6oYOzLVE_pQ2WHaL2szsMhavyX3wutXP3htHEVnyfmnJrfExoCuFsQAvD_BwE. Accessed 14 June 2022.
31. MADCAD. (2019). *What makes MADCAD.com unique?* Retrieved from https://www.madcad.com/tour/. Accessed 10 June 2022.
32. PlanRadar. (2021). *Digital management of construction tasks, communication and reports directly in the field*. Retrieved from https://www.planradar.com/sk/. Accessed 10 June 2022.
33. Fieldwire. (2022). *Jobsite managment for construction teams*. Retrieved from https://www.fieldwire.com/. Accessed 10 June 2022.
34. Procore. (2021). *Make it easier to manage construction*. Retrieved from https://www.procore.com/en-gb. Accessed 1 June 2022.
35. Google Play. (2021). *Buildo – Construction diary*. Retrieved from https://play.google.com/store/apps/details?id=net.spacive.buildo&hl=sk. Accessed 6 Janurary 2022.
36. Firstis. (2022). *BUILDARY ONLINE*. Retrieved from https://firstis.eu/sk/products/stavebny-dennik. Accessed 5 June 2022.
37. Google Play. (2022). *Asana*. Retrieved from https://play.google.com/store/apps/details?id=com.asana.app&hl=sk. Accessed 5 June 2022.
38. Google Play. (2022). *Wrike*. Retrieved from https://play.google.com/store/apps/details?id=com.wrike&hl=sk. Accessed 2 June 2022.

The Manufacturing Process of Die Forging Polyamide: Influence of the Annealing on the Properties and Structure

Waldemar Matysiak, Dorota Czarnecka-Komorowska, Lucia Knapčíková, and Jakub Hajkowski

1 Introduction

Nowadays, the industry is looking for innovative metallic or polymeric materials, including composites [1–4], as well as increasingly accurate and faster technologies for shaping products from these materials. Such opportunities are provided by the development of Industry 4.0 technologies [5–7]. The introduction of these innovations has been referred to as an industrial "revolution", as they have led to increased automation and robotization, productivity and efficiency of the production process, but also to a complete redefinition of the way of production, as well as work methods [5, 7].

Some technological processes for shaping metal or plastic products, such as forging, require improvement, in the direction of improving, the quality of the shaped products. Quality [8] in this case defined as the dimensional accuracy of forgings obtained by free forging technology is significantly lower than those obtained by die forging technology (in closed and open dies) (Fig. 1) [9, 10]. While the calculation of the closed-die forging volume and its preparation for this forging type must be very accurate because inaccuracies in the volume of the forging lead either to incomplete filling of the dies (if the volume of the forging is too small) or to the destruction of the tools (if the volume of the forging is too large).

Forging is a very important branch of metal-forming technology [10]. Forgings are used in many fields of technology, from simple fasteners (screws) to very

W. Matysiak (✉) · D. Czarnecka-Komorowska · J. Hajkowski
Poznan University of Technology, Poznan, Poland
e-mail: Waldemar.Matysiak@put.poznan.pl

L. Knapčíková
Department of Industrial Engineering and Informatics, Faculty of Manufacturing Technologies with a Seat in Prešov, Technical University of Košice, Prešov, Slovakia

L. Knapčíková, D. Peraković (eds.), *7th EAI International Conference on Management of Manufacturing Systems*, EAI/Springer Innovations in Communication and Computing, https://doi.org/10.1007/978-3-031-22719-6_5

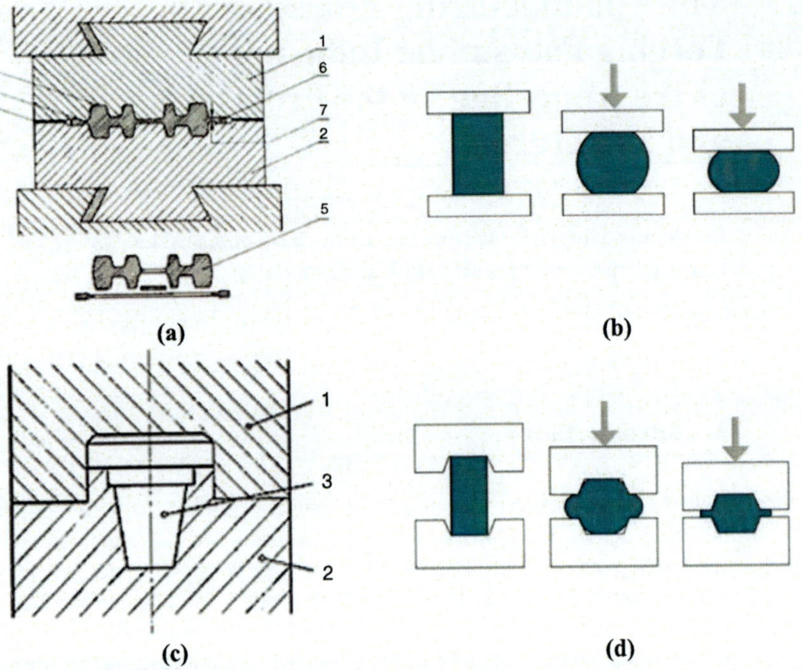

Fig. 1 Comparison of forging processes [9, 11 and own work]: **a-b**) open-die forging (1 – upper die, 2 – lower die, 3 – die cavity, 4 – parting line, 5 – workpiece, 6 – flash gutter, 7 – die land height, 8 – die land), **c-d**) closed-die forging (1 – upper die, 2 – lower die, 3 – workpiece)

complex parts of cars, airplanes, etc. [12]. The diversity of forging parts can be illustrated by the fact that, according to experts, there are more than 600,000 different forging tools in the world. The weight of forgings that can be obtained by forging ranges from 1 gram to more than 1 ton. The intensive development of the automotive industry has given new impetus to the development of forging technology [13]. Since this processing takes place above the recrystallization temperature, it is possible to forge almost all types of materials. The accuracy and quality of forged surfaces are at a relatively low level, so the usable surfaces of the forging must then be machined. In contrast, the mechanical properties of forgings are very good, so responsible parts, such as engine pistons, vehicle axles, etc., are made almost exclusively by forging. On the other hand, the shape (geometry) of the die cavity depends on the shape of the forging.

In the literature [14], you can find specific rules for the construction of forgings with all their elements (thickness of bridges, split planes, internal and external rounding, etc.). Of particular importance is the shape of the bridge and the stock forging. As mentioned earlier, during the forging process in open dies, efflorescence (excess material) is formed, which must be removed after the forging process. This is done with special trimming tools. It should be noted that such tools consist of a die and a punch, the shape of which is adapted to the shape of the forging. After the hot forging process, it is possible to subject the forging to calibration. This is a procedure in which compressive stresses are applied to specific surfaces of the forging using tools in the form of flat plates. This operation is performed cold and is intended to improve the quality and accuracy on certain surfaces of the forging [8]. The surfaces after calibration have a quality that can be compared with that obtained by machining operations. Calibration is also used when it is necessary to achieve accuracy in forging in terms of parallelism between certain surfaces.

Modern forging manufacturing processes are aimed at giving the required shape and obtaining the appropriate mechanical properties of the product [11]. Currently, to shape the mechanical properties of forgings, the Polish forging industry uses traditional technologies, which include heating and quenching of the charge, die forging, and the application of heat treatment with hardening and tempering operations. For forgings for the automotive industry, high mechanical properties with good ductility are required due to high dynamic and static loads during vehicle operation. The directions of forging technology development are now mainly focused on obtaining forgings without technological allowances and on reducing their manufacturing costs. For this reason, since the early 1990s, forgings with a ferritic-pearlitic structure, strengthened by separating vanadium, niobium, and titanium micro-additives, have been increasingly used in the automotive industry. Shaping of the microstructure and mechanical properties of these forgings is done by forging combined with controlled cooling, known as thermomechanical forging. In less responsible applications, the desired mechanical properties of the forgings can be achieved without the use of alloying micro-additives, controlling only the temperature of the material during forging and the cooling rate after forging. Compared to the traditional technology for manufacturing forgings with high mechanical properties involving additional heat treatment, the use of thermomechanical forging offers certain advantages, such as eliminating costly and environmentally unfriendly traditional heat treatment, reducing the number of process operations, shortening manufacturing time, reducing manufacturing, storage, and transportation costs, and increasing competitiveness by lowering manufacturing costs.

Due to the analogy of the process and the physical phenomena occurring in polymeric materials, the authors of the chapter decided to apply the technology of die forging in closed dies with an outflow to shape products made of semicrystalline plastics, i.e., polyamide, polyoxymethylene, and polybutene. For example, in a patent, a process for the forging of rigid, crystallized plastic was described [16]. Die forging creates a certain other direction in the manufacturing processes of plastic products [15].

Polyamide is one of the semicrystalline polymeric materials suitable for the production of engineering products, whose important feature of the structure is the presence of amorphous (disordered) areas in addition to ordered crystalline areas, and a measure of the degree of order is the degree of crystallinity of the polymer, which depends on the degree of secondary annealing and the rate of cooling of the polymer, which significantly affects the properties and structure of polyamide.

The purpose of this study was to develop a die forging process for polyamide forgings, along with evaluating the effect of annealing temperature on the structure and physical properties of polyamide (PA).

2 Methodology

2.1 Material and Manufacturing of PA Samples by Forging

The semicrystalline polyamide (PA) with a density of 1.113 g/cm^3 and a hardness of about 80 N/mm^2 for machining was tested.

In the die forging process, new technologies are being developed to minimize the number of intermediate operations. In the case of the forging of polyamide forging, three forging operations were performed each time. The final decision is made with a cost balance that takes into account several other factors, including forging yield, tool life, and the size of the machine used to forge the matrix product. Industrial practice shows that for elongated forgings with a significant difference in cross-sectional area, the use of a rolled pre-forging increases the efficiency of the process by an average of two times, the yield increases up to 20%, tool life increases by 50%, while the labor cost decreases by about 15%. Other parameters of this technology, including the mechanical properties of the products, can be predicted by considering the distribution of deformations and temperature changes with the duration of the forging cycle.

In the case of the PA forging in question, the duration of the entire forging process was 10 sec. In the first stage of shaping in the turning process, pre-forging was prepared (Fig. 2). The forging process was carried out for the forged forgings, which were annealed to160 °C. The forged forgings were annealed in a Nabertherm L3/11s furnace, for 30 min. Forging was carried out using a PC100 friction-screw press.

The presented forging was purposely chosen for the study because it has the characteristic shape of an elongated forging with two thickened ribs with a maximum cross-sectional area several times larger than the minimum cross-sectional area. The ribs stiffening the stem of the forging are difficult to shape, which sometimes requires the use of a pre-die blank. The elimination of the pre-die blank when forging such a forging proved possible, as a result of the use of a strictly defined shape of the pre-forging with variable diameters.

The forging, hatchet-shaped, was made of cast polyamide (PA). Figure 2 shows the geometry of the forging (a) made in Autocad 2022 Edu, a front view of the forging (b), the upper die-workpiece (c), and the finished product (d), and the forging preform (e).

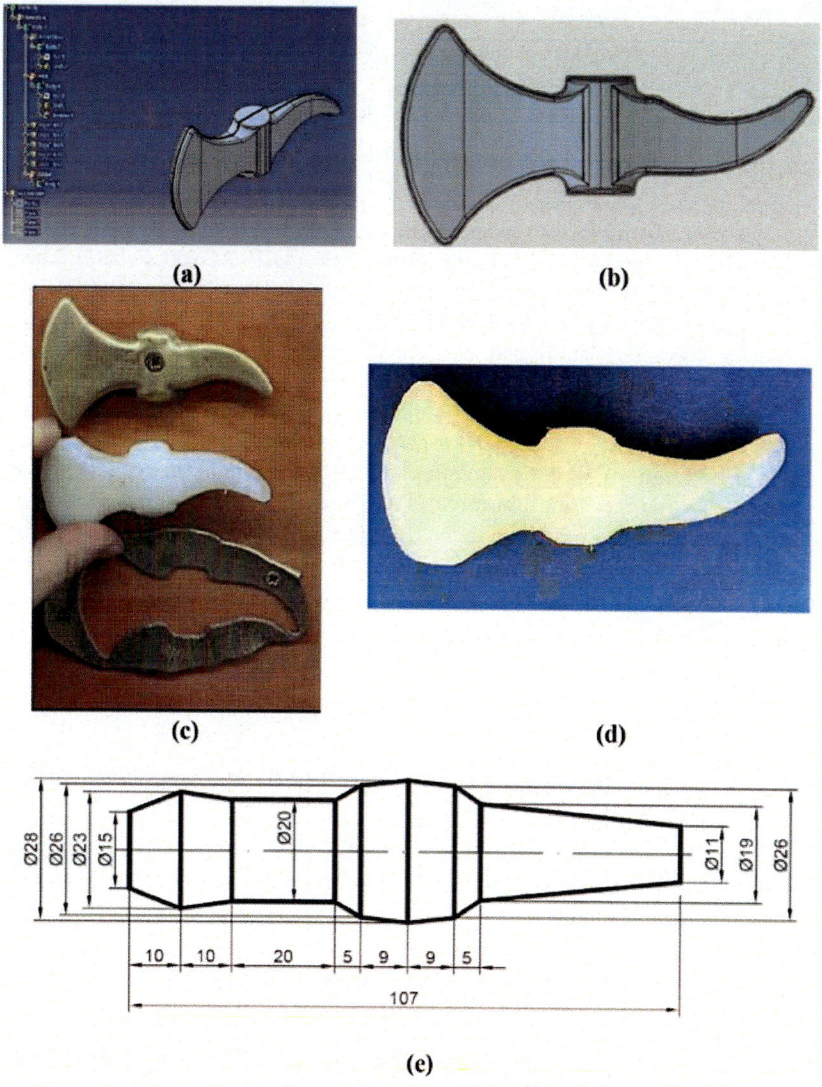

Fig. 2 A view of hatchet (own work): (**a**) geometry of forging in AutoCAD 2022 Edu software, (**b**) front view, (**c**) upper die-workpiece, (**d**) finished product, (**e**) forging preform

2.2 Annealing Process

Polyamide (PA) samples were subjected to annealing at 130 °C, 140 °C, 150 °C, and 160 °C for 4 h. The polyamide was then conditioned in water for 24 h. An unannealed, reference polyamide was dried in a vacuum with no heat for at least 24 h to remove water residue.

2.3 Methods Characteristic

The density of the PA was measured using the hydrostatic method according to ISO 1183-1: 2019 [17], using an electronic scale (AXIS AD200, Poland). The hardness of the samples was measured using a ball hardness Durometer (KB Prüftechnik Inst.), following the ISO 2039:2001 [18] procedure. The thermal behavior of the polyamide before and after annealing was performed using the DSC 204 F1 Phoenix apparatus (NETZSCH GmbH, Germany) operating with a nitrogen flow of 20 mL min^{-1} as a protective atmosphere. Samples were first heated at a rate of 10 °C min^{-1} from 23 °C to 270 °C temperature and kept at this temperature for 12 min followed by cooling to 23 °C at the rate of 10 °C min^{-1} to eliminate the thermo-mechanical history of PA. The thermally treated samples in their closed aluminum pans were used in the DSC measurements. DSC studies were conducted to determine the melting (T_m) and crystallization temperature (T_c), melting enthalpy (ΔH_m), and degree of crystallinity (X_c). The X_c of the PA was calculated using the following Eq. (1):

$$X_c\,(\%) = \frac{\Delta H_m}{\Delta H_m^0} \bullet 100 \tag{1}$$

where ΔH_m is the melting enthalpy of the sample [Jg^{-1}], ΔH_m^0=239 Jg^{-1} the melting enthalpy for a 100% crystalline polyamide [19].

3 Results and Discussion

3.1 Hardness and Density Results Analysis

The results of the hardness tests and density on the surface of the forgings under various heat treatment conditions are shown in Fig. 3.

The hardness measurements were carried out on both sides of the die forging, i.e., from the forging and the opposite side. As shown in Fig. 3a, the hardness of the unannealed sample was higher about 14% for the forging surface than for the opposite of the surface. This is due to the course of the forging process, which causes surface hardening of the polyamide layer, as a result of secondary crystallization. Heating at 130–160 °C, however, polyamide before the forging process facilitates the process of shaping the forging and improves the surface quality of the product.

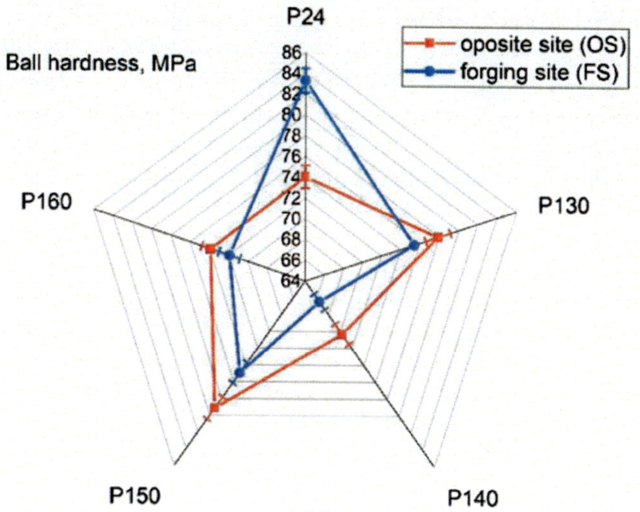

(a)

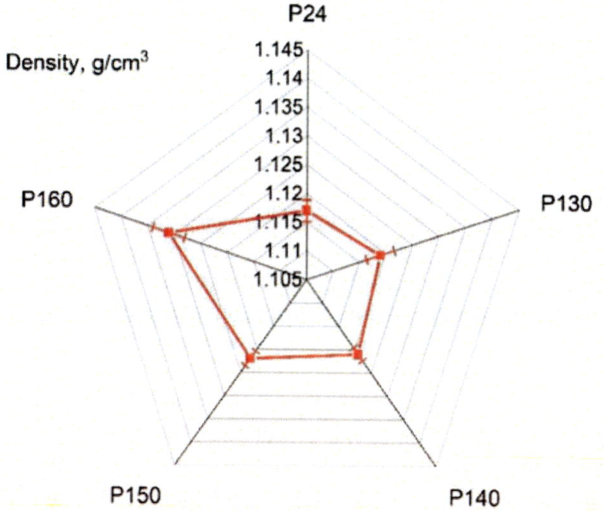

(b)

Fig. 3 The hardness **(a)** and density **(b)** of PA samples before and after various thermal treatments (own work)

The higher value of hardness of polyamide might be due to the smaller spherulite size during the forging process and an increase in the packing density of macromolecules. This is under the density results obtained (Fig. 3b) for polyamide. Density testing conducted by hydrostatic method for polyamide forged at 160 °C

Table 1 DSC data for polyamide (PA) after the annealing process

Sample code	T_{m1}[°C]	T_{m2}[°C]	ΔH_{m2}[J/g]	T_{c1}[°C]	T_{c2}[°C]	ΔH_{m2}[J/g]
PA 24T	222.4	219.1	−51.3	167.0	169.3	201.5
PA130T	219.7	218.4	−48.1	168.8	169.6	176.9
PA140T	221.6	218.2	−41.9	169.3	169.0	151.4
PA150T	219.9	218.0	−46.7	17.5	166.3	133.2
PA160T	222.0	217.1	−59.3	170.0	167.0	139.7

(around the crystallization temperature of PA) showed a slight increase in density compared to unforged polyamide, which is due to an increase in the degree of crystallinity of the polyamide. As forging cycles increase, fragmentation of the spherulites is observed, which have little resistance to force and are mostly disintegrated by lamellae regrouping.

3.2 Structure Analysis

The influence of annealing temperature on the structure of polyamide (PA) forgings based on DSC tests was investigated. The DSC results are shown in Table 1 and the melting (a-b) and crystallization curves (c) are shown in Fig. 4. The X_c results for PA after the annealing are illustrated in Fig. 5.

The enthalpy of melting (ΔH_m), melting (T_m) and crystallization temperature (T_c), and degree of crystallinity (X_c) were evaluated. The X_c was calculated by integrating the area under the DSC melting endothermic peak and dividing by the heat of fusion with 100% crystalline PA, according to Eq. 1.

The DSC thermograms (Fig. 4a, b) show only single melting entoderms with a peak temperature of $T_m = 220$ °C, suggesting the melting of a single type of crystallites with the same phase crystal form having the same crystal thickness. Kyotani and Mitsuhashi [20] suggested that this phenomenon was related to the occurrence of the melting of α-crystals. DSC results showed insignificant changes in the melting temperature T_m (about 222 °C) in the first heating cycle and the second heating cycle.

On the other hand, at an annealing temperature of 160 °C for polyamide, an increase in the crystallization temperature of 3 °C (to 170 °C) was observed for the first heating cycle. It should be noted (Fig. 5) that the X_c values of PA160 at $T = 160$ °C are 15% higher than that of the PA reference sample, which explains the occurrence in polyamide of the phenomenon of recrystallization, which near the melting point of polyamide causes an increase in crystallites, that is, leads to the formation of a coarse-grained structure. Recrystallization in polyamide is manifested by an increase in the parameters of the crystal lattice caused by the sliding motion of the corrugated polymer macromolecules. The second effect of annealing can be grain growth. This is under previous work by Koszkul et al.

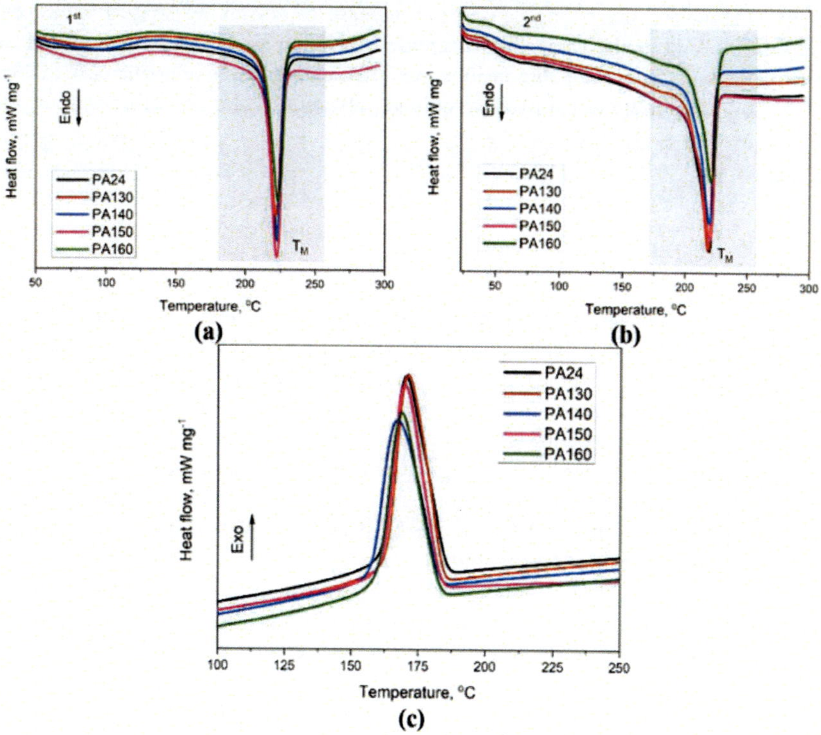

Fig. 4 DSC curves of PA before and after various thermal treatments (own work): (**a**) first heating traces, (**b**) second heating traces, and (**c**) second cooling (crystallization)

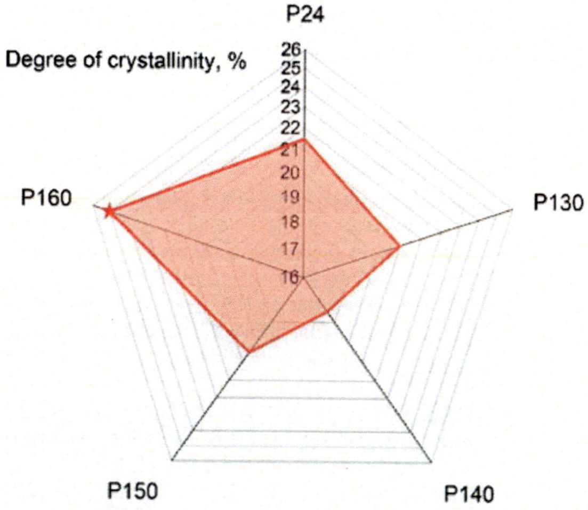

Fig. 5 The degree of crystallinity for PA samples before and after various thermal treatments (own work)

[21], where annealing caused the increase of crystalline phase fracture by about 20% and at the same time decreased of amorphous phase fraction and also the increased spherulite dimensions. A similar phenomenon of an increase in the degree of crystallinity of halloysite-modified polyamide under annealing was described by Mencel et al. [22].

4 Conclusions

In this chapter, a die forging process was developed in a closed mold of a semicrystalline polyamide forging. The tooling necessary to carry out the forging was designed, and the process parameters were defined. In the first stage of shaping in the turning process, the forgings were prepared. For the polyamide forgings, the duration of the entire forging process was 10 seconds. The forging process was carried out for the pre-forging, which was annealed at 130 °C to 160 °C. The developed process of manufacturing forgings was aimed at giving the required shape and obtaining the appropriate mechanical and structural properties of the plastic products. For this purpose, the shaped polyamide forgings after annealing to 160 °C and crystallization in water were subjected to evaluation of changes in selected physical properties of polyamide forgings [23, 24]. It is shown that polyamide forgings subjected to forging at room temperature show higher hardness on the forging side than on the opposite side, which is due to a change in the packing density of macro-molecules in the main chain. DSC results showed that at the higher annealing temperature for PA160 there is a marked increase in hardness and degree of crystallinity, due to secondary recrystallization of the polyamide [20]. In the future, research should focus on forging other plastic materials at different temperatures and heating times.

References

1. Sika, R., Rogalewicz, M., Popielarski, P., Czarnecka-Komorowska, D., Przestacki, D., Gawdzińska, K., & Szymański, P. (2020). Decision support system in the field of defects assessment in the metal matrix composites castings. *Materials, 13*, 3552. https://doi.org/10.3390/ma13163552
2. Czarnecka-Komorowska, D., Sterzynski, T., & Dutkiewicz, M. (2016). Polyoxymethylene/polyhedral Oligomeric Silsesquioxane composites: Processing, crystallization, morphology and thermo-mechanical behavior. *International Polymer Processing, 31*, 598–606. https://doi.org/10.3139/217.3243
3. Knapcikova, L., Duplakova, D., Radchenko, S., & Hatala, M. (2017). Rheological behavior modelling of composite materials used in engineering industry. *TEM Technology, Education, Management, Informatics, 6*, 242–245. https://doi.org/10.18421/TEM62-07
4. Czarnecka-Komorowska, D., & Mencel, K. (2014). Modification of polyamide 6 and polyoxymethylene with [3-(2-aminoethyl)amino]propyl-heptaisobutyl-polysilsesquioxane nanoparticles. *Przemysł Chemiczny?, 93*, 392–396. https://doi.org/10.12916/przemchem.2014.392

5. Przemysł 4.0 | Definicja, technologie, korzyści | SAP Insights Available online: https://www.sap.com/poland/insights/what-is-industry-4-0.html. Accessed on 21 June 2022.
6. Ciszak, O. (2018). Industry 4.0 – Industrial robots. In *Proceedings of the proceedings of the 3rd EAI international conference on management of manufacturing systems*. EAI.
7. Ciszak, O., Juszkiewicz, J., & Suszyński, M. (2020). Programming of industrial robots using the recognition of geometric signs in flexible welding process. *Symmetry, 12*, 1429. https://doi.org/10.3390/sym12091429
8. How to identify the quality of Forging Available online: https://shdhforging.com:443/news/how-to-identify-the-quality-of-forging. Accessed on 21 June 2022.
9. Matysiak, W., & Plancak, M. (2021). *TERMINOLOGIA OBRÓBKI PLASTYCZNEJ*. Wyd. Politechniki Poznańskiej.
10. ASM Metals Handbook, Vol 14.Pdf Available online: https://docs.google.com/file/d/0B3Mrzn2Z7-tSWVd0OXk4ZXd0Q0U/edit?usp=embed_facebook. Accessed on 21 June 2022.
11. Gupta, A. (2018). What are the advantages and disadvantages of closed die forging? *Medium*.
12. Closed Die Forging and Open Die Forging - Buy Die Forging, Die Forging Part, Closed Die Forging Product on Qingdao Haozhifeng Machinery Co., Ltd Available online: http://www.haozhifeng.com/Closed-Die-Forging-and-Open-Die-Forging-pd543918.html. Accessed on 21 June 2022.
13. Metal Forging Processes, Methods, and Applications | TFGUSA Available online: https://www.tfgusa.com/metal-forging-processes-methods/. Accessed on 21 June 2022.
14. ASM HandBook Volume 14 - Forming and Forging [PDF] | Online Book Share Available online: https://docero.tips/doc/1942-asm-handbook-volume-14-forming-and-forging-xgjpogypz2. Accessed on 21 June 2022.
15. Tay, C. Y., Hu, X., & Liu, B. (2020). Forging new frontiers in polymer research and innovations. *Macromolecular Rapid Communications, 41*, 2000521. https://doi.org/10.1002/marc.202000521
16. Kulkarni, K. (1974). Forging of rigid crystalline plastics.
17. Plastics — Methods for determining the density of non-cellular plastics — Part 1: Immersion method, liquid pycnometer method, and titration method. Polish Committee for Standardization, Poland, 2013. ISO 1183-1:2019 Available online: https://www.iso.org/cms/render/live/en/sites/isoorg/contents/data/standard/07/49/74990.html. Accessed on 19 May 2022.
18. Plastics — Determination of hardness — Part 1: Ball indentation method ISO 2039-1:2001 Available online: https://www.iso.org/cms/render/live/en/sites/isoorg/contents/data/standard/03/12/31264.html. Accessed on 26 April 2021.
19. Millot, C., Fillot, L.-A., Lame, O., Sotta, P., & Seguela, R. (2015). Assessment of Polyamide-6 crystallinity by DSC. *Journal of Thermal Analysis and Calorimetry, 122*, 307–314. https://doi.org/10.1007/s10973-015-4670-5
20. Kyotani, M., & Mitsuhashi, S. (1972). Studies on crystalline forms of Nylon 6. II. Crystallization from the melt. *Journal of Polymer Science Part A-2: Polymer Physics, 10*, 1497–1508. https://doi.org/10.1002/pol.1972.160100807
21. Koszkul, J., & Kwiatkowski, D. (2006). The influence of annealing on dynamical mechanical properties of polyamide 6 / fiber glass composites. *Journal of Achievements in Materials and Manufacturing Engineering, 19*, 5.
22. Mencel, K., & Kemnitz, Ł. (2019). Influence of heat treatment on the structure and properties of polyamide 6/ Halloysite nanocomposites. *Archives of Mechanical Technology Material, 39*. https://doi.org/10.2478/amtm-2019-0002
23. Czarnecka-Komorowska, D., et al. (2021). Polyethylene/polyamide blends made of waste with compatibilizer: Processing, morphology, rheological and thermo-mechanical behavior. *Polymers, 13*(14), 2385. https://doi.org/10.3390/polym13142385
24. Krawiec, P., Wargula, L., Maloziec, D., Kaczmarzyk, P., Dziechciarz, A., & Czarnecka-Komorowska, D. (2020). The toxicological testing and thermal decomposition of drive and transport belts made of thermoplastic multilayer polymer materials. *Polymers, 12*, 2232. https://doi.org/10.3390/polym12102232

Part II
Sustainable Communications & Information and Communications Technology

Application of the SMART Optical Displacement Sensors in Measuring the Diameter Deviation of C45 Steel After the Turning Process

Martin Miškiv-Pavlík, Jozef Jurko, Alexander Hošovský, Vratislav Hladký, Kamil Židek, Igor Petruška, and Tadeusz Eugeniusz Zaborowski

1 Introduction

As industrial production is increasingly changing from manual labor to highly automated and demanding work, the application of modern measurement technology is a main part of this production. One of the most important tasks in the technical production industry is the inspection of products. Devices with contact measurement methods of the obtained parameters are still used for quality control of machined surfaces. In these contact measuring methods, the measuring contact is in direct contact with the machined surface. This direct contact of the measuring contact with the machined surface is one of the negative effects of the measurement. On the one hand, the contact surface of the measuring device is damaged, and on the other hand, the surface to be measured is damaged. One of the many possible uses of contactless sensors is to measure deviation of surface from workpiece diameter after process of turning. This method allows rapid detection of defects on the workpiece surface without sending the product to quality control after production [1–8].

The chapter main thesis was the application of optical contactless sensors in the measurement of the deviation parameter from the workpiece diameter of C45

M. Miškiv-Pavlík · J. Jurko (✉) · A. Hošovský · V. Hladký · K. Židek
Technical University of Košice, Faculty of Manufacturing Technologies, Prešov, Slovakia
e-mail: martin.miskiv-pavlik@tuke.sk; jozef.jurko@tuke.sk; alexander.hosovsky@tuke.sk; vratislav.hladky@tuke.sk; kamil.zidek@tuke.sk

I. Petruška
University of Prešov in Prešov, UNIPO, Prešov, Slovakia
e-mail: igor.petruska@unipo.sk

T. E. Zaborowski
Instytut Badań i Ekspertyz Naukowych, Gorzów Wielkopolski, Poland
e-mail: tazab@sukurs2.pl

L. Knapčíková, D. Peraković (eds.), *7th EAI International Conference on Management of Manufacturing Systems*, EAI/Springer Innovations in Communication and Computing, https://doi.org/10.1007/978-3-031-22719-6_6

steel after the turning process. As a comparison to optical sensors, a mechanical micrometer was used in the experiment as a standard for the measured values. In this experiment, we demonstrated the feasibility of using these sensors to directly detect surface damage and other undesirable features of processed materials in processing centers, eliminating the need for post-processing quality control increase.

2 Optical Sensors Measuring Literature Review

In contrast to contactless testing, many traditional distance measuring devices such as micrometers and calipers make contact with the workpiece surface. This method involves contact that damages not only the instrument but also the object being measured [9].

If the object to be measured is not fixed or intended to move, the measurement results will be inherently unstable and inconsistent on subsequent measurements. The current market for sensors is growing dynamically. With over 1000 different sensor types in different countries, choosing the right sensor type for a particular application can be difficult. Machining technology is realized in the system machine, tool, and workpiece. In relation to the machine, Kishore et al. applied optical sensors vibration detection sensors to monitor the life of a diesel engine, based on the amount of measured vibrations (distance deviations) it produced [10, 11, 12] (Fig. 1).

Sensors are defined as units capable of detecting errors to defined principles. The head of sensor as the key part of the device is the interface between the external stimulus and the internal components for processing and further analysis

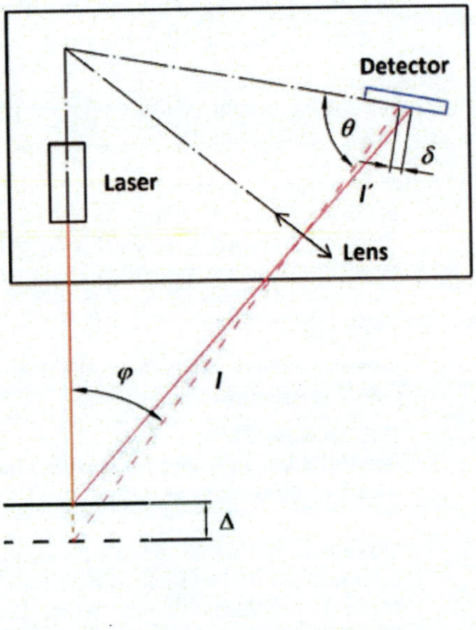

Fig. 1 Laser displacement triangulation sensor scheme [12]

of the information measured during the application [13, 14]. The main principle of this measuring method is application of polychromatic white light. This light is projected from the sensor head through a multi-lens onto the workpiece surface. The so-called confocal lens arrangement is used, in which the light is divided into colored components with different wavelengths by the natural diffraction of light. The reflected radiation to the confocal aperture of the sensor, through which only focused radiation of a specific wavelength reaches the sensor. The light quantity returned to the optical receiver changes depending on the position and distance of the workpiece. In a study, Zou et al. describe the application of non-contact sensors using a neural network in the detection of points on the surface of the workpiece, on the basis of which they have developed an active control of machining during the process. Based on the elimination of spindle deflection and accurate measurement of the center axis of the workpiece, they were able to create a 3D model of the measured workpiece and compare it with the resulting workpiece after the process of turning. In this way, they were able to detect the error in serial processing. Yu et al. used a chromatic confocal sensor in their study to measure the unevenness (thickness) of transparent materials. This study showed that with proper illumination, it is possible to use confocal sensors to measure not only metallic materials, but also materials such as glass and liquids with a measurement error rate of only 2% [15–19].

This type of sensor can be used to optimize, control and diagnose production reliability based on workpiece features on the surface. Laser scanners are ideal for these applications due to their high speed, repeatability, and accuracy of up to 1 micron. Fu et al. designed a measurement scheme using two confocal sensors facing each other, with the help of which he measured the topographic changes of the surface of the machined material directly after the machining process. The measurement system was implemented in a robotic arm, and the distance and angle measurements were adjusted during the measurement based on the computer-aided design model of the respective component. Experimental data confirm the ability of this system to measure surface roughness in the range Ra 0.002–0.007 mm ($\lambda c/\lambda s$ 300, accuracy was 5%). Cheng et al. used a monochromatic confocal sensor in the measuring process of the workpiece surface with the method of a straight line along the surface, creating a topographic map of the machined product. Measured data proved that the proposed system can achieve accuracy comparable to mechanical laboratory conditions. The measurement error is 0.001 mm. Since the current measuring mechanisms are mainly used in development processes, the present chapter can be used for new applications [20–23] (Fig. 2).

3 Research Methodology

As mentioned in the previous chapter, two types of optical deviation sensors were used in the experiment, both of which operate on the same principle, namely the insensitivity of the light reflected from the workpiece back to the sensor. Although

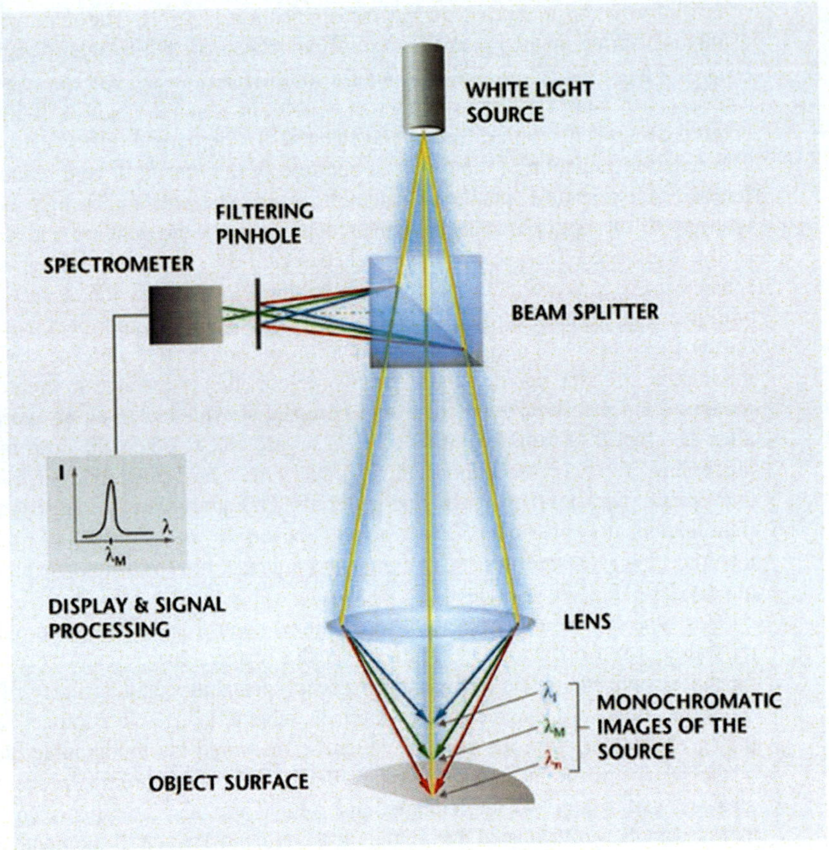

Fig. 2 Principle of confocal measuring sensor [23]

the principle of these technologies is almost the same, the circuit diagram for these sensors is not.

3.1 Confocal Sensor

In the case of confocal (CS) sensor, there is no need to application of additional hardware components. The data are written directly into the software database, which is the biggest advantage in comparison to other measuring devices that require the manual input of measured data. With the increasing distance from the sensor to the measuring workpiece, the inaccuracy of the measurement also increases, what presents the biggest issue of these types of sensors. Besides the limitation of distance

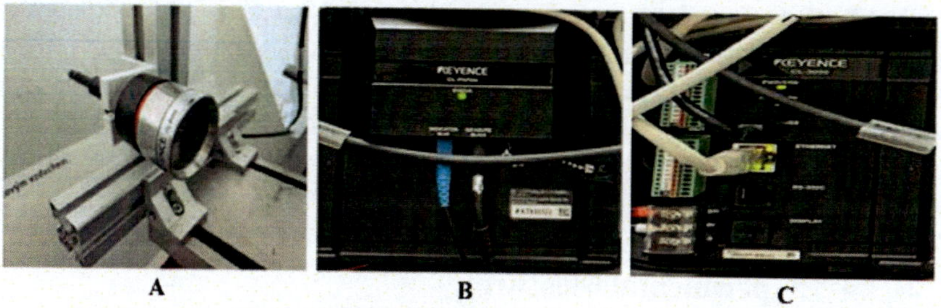

Fig. 3 Measuring scheme of confocal sensor (CS)

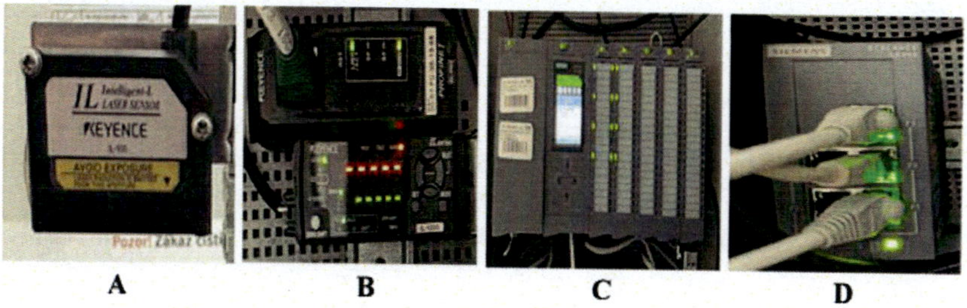

Fig. 4 Measuring scheme of triangulation sensor (LTS)

in the measuring process, other disadvantages are see-through materials or glossy types of surfaces, which affect the measurement accuracy.

From Fig. 3 the wiring scheme of the measurement system can be briefly seen. Applied measuring system consists of the measurement head of the Keyence CL – P070 (A) with sensor installed on the rotary holder of the (Leadwell-T5) cutting tool, to achieve a dynamic type of measurement for the machined workpiece. All digitized values of the sensor head were transmitted by means of an amplifier Keyence CL – 3000 (C) to the communication module Keyence CL – P070N (B), where microscopic irregularities of the workpiece diameter could be observed via Ethernet to the software (Navigator).

3.2 *Laser Triangulation Sensor*

In the experiment part, we prepared a measurement system, which scheme is shown in Fig. 4.

The measurement system consists of measuring head (A) Keyence IL100. Its measuring range is 70–130 mm, measuring spot diameter (400 × 1350) μm which depend on the workpiece distance, repeatability 4 μm, linearity ±20 mm, and weight 75 g.

The communication unit (B), DL – PN 1. These elements are connected to the sensor head via an extension cable and are connected to the PLC (C), Siemens 1511C via the profinet module. The switch unit (D), Siemens XB005 is used as a boundary (connection) via ethernet between the PLC and the evaluated unit (PC–TIA Portal software).

4 Experimental Measuring

The first objective of this chapter was to test the possibility of using optical displacement sensors to measure the deviation of the diameter of round C45 steel machined under the same turning conditions but with another measuring devices (CS, LTS) (Fig. 5).

The basic parameters of the experimental workpiece before machining were diameter $d = 14.99$ mm and length $l = 88.8$ mm.

The machining conditions in the turning process were feed (0,28 mm), cutting speed (60 m/min), and cutting depth (0,1 mm) (Fig. 6).

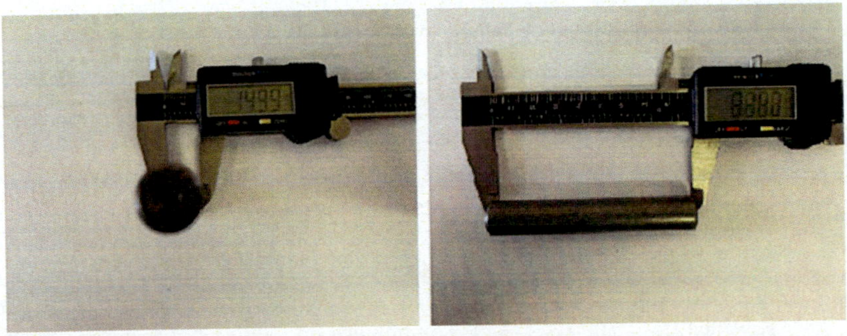

Fig. 5 Diameter measuring of the C45 sample

Fig. 6 Figure of the workpiece working zone during the turning process

A B

Fig. 7 Measuring of the diameter deviation by ((**a**) LTS, (**b**) CS)

Fig. 8 Measuring of the diameter with LIEVE micrometer

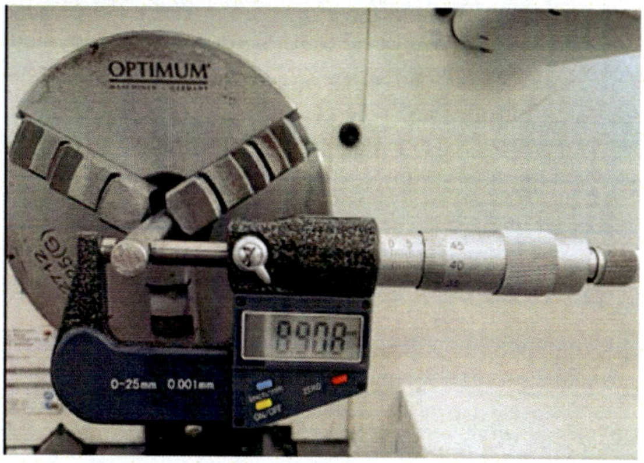

After the machining, the first TLS was inserted into the tool holder, which we aligned exactly with the center of the workpiece and then defined the zero point at the base of its measuring range (100 mm). In the next step, we divided the sample into 6 sections of 10 mm in length and a circumference of 72 degrees and measured the deviation from the diameter at each section (Fig. 7).

The same measurement principle was used at CS when we reinserted the sensor into the tool holder, set the zero point for CS to (70 mm), and then measured the deviation values from the workpiece diameter. For the purposes of this experiment measuring holders were constructed for both TLS and CS sensors using a 3D printing with Filalab PLA filament.

A LIEVE micrometer with 0–25 mm measuring range and an accuracy of 1 μm were used in this experiment as a standard for the measured values (Fig. 8).

4.1 Experimental Results

Since it was necessary to create custom measurement software for the LTS, the data measured by the LTS was written by the PLC to a text block in the form of measurement time, measurement value, and write index. During the measurement sections, 200 values were measured, from which an arithmetic mean was formed. All data were subsequently processed from text block to Excel program for further analysis (Fig. 9).

In the case of CS, data processing was somewhat easier because Keyence also provides the sensor with software to analyze and process the measurement data, so it was possible to measure more values with one confocal sensor, exactly 1000 measurements per second with an accuracy of 1 μm. The measurement data can be evaluated directly in the CLNavigatorN-N program or exported to Excel (Fig. 10).

As mentioned in the previous part, measured values from the sensor head were saved from the software or PLC to a table in Excel. In Table 1, data of LTS in the form of arithmetic mean values are processed, which were determined from all values for a certain section. In Table 2, data from the same workpiece are processed, but using CS. The last data in Table 3 were processed data measured from micrometer measurements.

The measuring system determines only the zero position deviation, for the real values the following equation must be applied:

$$\Delta_d = l + x \pm y \tag{1}$$

where

Δ_d – real deviation of diameter,
l – measured deviation,
x – the required value,
y – sensor error.

Fig. 9 Example of measured values from LTS

```
00:00.031;55,185;12,336;12336
00:00.046;55,181;12,336;12336
00:00.046;55,169;12,336;12336
00:00.046;55,166;12,336;12336
00:00.046;55,162;12,336;12336
00:00.046;55,158;12,336;12336
00:00.046;55,151;12,336;12336
00:00.062;55,144;12,336;12336
00:00.062;55,139;12,336;12336
00:00.062;55,135;12,336;12336
00:00.062;55,129;12,336;12336
00:00.062;55,123;12,336;12336
```

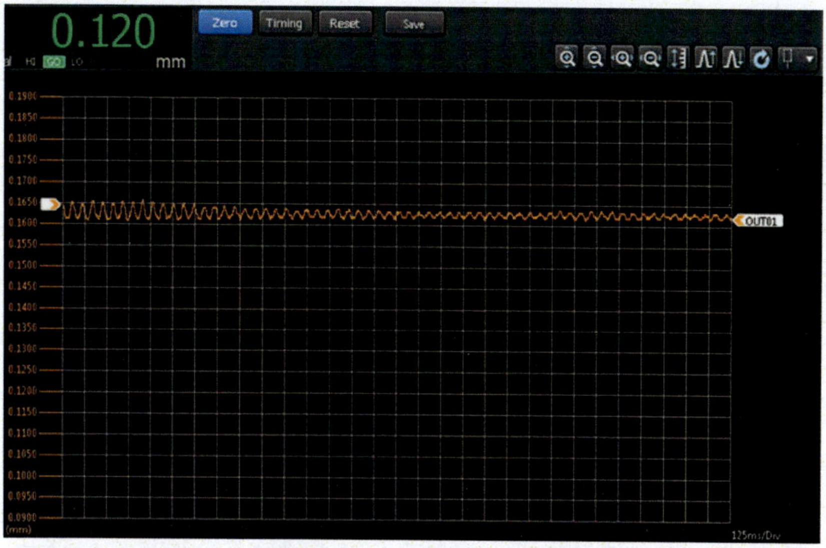

Fig. 10 Example of the measured data from CS in CLNavigatorN-N software

Table 1 Measured data for LTS

Distance	Measured angle of workpiece (°) and deviation in (mm)				
(mm)	0/360°	72°	144°	216°	290°
0	−0.041	−0.069	−0.177	0.099	0.091
10	−0.049	0.002	−0.152	0.034	0.105
20	0.01	0.021	−0.078	0.027	−0.01
30	−0.111	−0.094	−0.022	0.093	**−0.18**
40	−0.03	−0.073	−0.141	0.042	0.089
50	−0.016	−0.021	−0.129	0.036	−0.009

Calculation example of the real measured dimension:

- x: 10 mm,
- l: 0.005 mm,
- y: 0.003 mm,

$\Delta_d = 10{,}005 \pm 0{,}003$ mm.

When measuring the workpiece with both sensors the lowest deviation was about - 0.18 mm and the highest deviation was 0.01 mm, due to the machined surface quality after turning.

From Fig. 11, we can see that the CS is a very accurate representation of the readings from the mechanical micrometer. The largest difference occurred in the 30 mm section, when the difference in the average of the measured values was 0.019 mm. An even greater difference between the LTS readings occurred when

Table 2 Measured data for CS

Distance	Measured angle of workpiece (°) and deviation in (mm)				
(mm)	0/360°	72°	144°	216°	290°
0	−0.041	−0.04	−0.044	0.018	0.015
10	0.03	0.009	−0.018	0.022	0.062
20	0.022	0.029	−0.057	0.088	0.071
30	0.037	0.076	−0.021	0.087	0.071
40	0.032	0.099	−0.026	0.033	0.075
50	0.064	0.026	−0.04	0.011	−0.055

Table 3 Measured data for micrometer

Distance	Measured angle of workpiece (°) and deviation in (mm)				
(mm)	0/360°	72°	144°	216°	290°
0	−0.035	−0.03	−0.039	0.021	0.03
10	0.019	0.096	−0.047	0.001	0.084
20	0.037	0.005	−0.027	0.076	0.081
30	0.025	−0.005	0.052	0.066	−0.02
40	0.02	**0.1**	0.039	0.019	0.11
50	0.063	0.022	0.009	0.008	−0.074

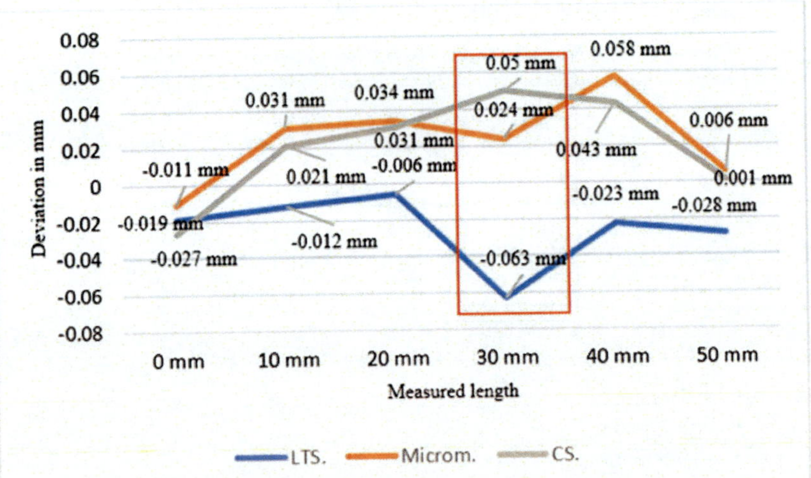

Fig. 11 Graphical comparison of individual measuring systems by deviation of diameter on length

the difference between the mechanical gauge and the LTS exceeded the difference by 0.039 mm, again in the 30 mm section. Regarding the measured values of the circumference of the workpiece from Fig. 12, we can again confirm that CS measured almost identical values as a mechanical gauge, the difference between a

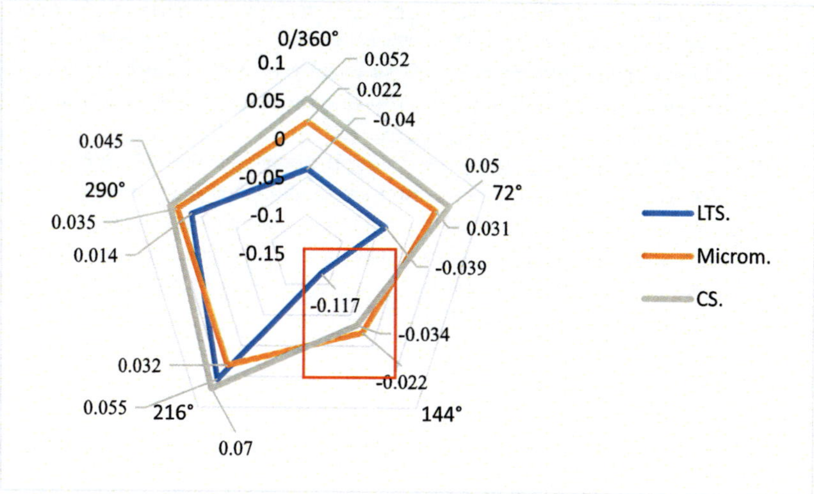

Fig. 12 Graphical comparison of individual measuring systems by deviation of diameter on workpiece angle

mechanical gauge and LTS increased on the contrary. At an angular rotation of 144°, the difference of the measured values reached up to 0.095 mm.

5 Discussion

In this chapter, a new experimental measurement system focused on contactless optical sensors was presented. Analyzing the diagrams, we can see that the average total deviation of the workpiece diameter is between - 0.18 mm and 0.01 mm. Based on these deviations, we can determine the surface errors of the machined material by identifying defectively machined workpieces in turning process. However, these measuring systems also have their limits, namely the measuring distance.

6 Conclusion

From the results of the experiment, it can be concluded that the confocal sensor copied the micrometer measurement data very accurately, with the highest deviation from the measured value being 0.019 mm. This is in contrast to the triangulation sensor, whose deviation from the diameter of the workpiece in comparison with the mechanical micrometer was at one point up to 0.095 mm. With this experiment we have clearly proved that the use of optical sensors has its application in the machine industry, especially in the measurement of surface properties of workpieces after turning.

The second part of research at the Faculty of Manufacturing Technologies in Prešov is the practical implementation of optical sensors in manufacturing operations as a substitute for post-machining quality control. Contactless sensors are one of the best technologies in the field of optical distance and position measurement. A disadvantage of small and medium-sized automated plants is the high purchase price of the device compared to the more common diffuse sensors or single-channel optical sensors, but in large-scale production and advanced production with automation processes, where they belong, where very high measurement accuracy must be achieved, these sensors are essential.

Acknowledgments All authors of the chapter would like to thank the research team of grant agency for the support of research works and the co-financing by the VEGA 1/0268/22 project.

References

1. Mrkvica, I., Neslušan, M., Čep, R., & Sléha, V. (2016). Properties and comparison of PVD coatings. *Tehnički vjesnik/Technical Gazette, 23*(2), 569–574. ISSN 1330-3651.
2. Petrů, J., Zlámal, T., Mrkvica, I., & Čep, R. (2013). The effect of thermal shocks on wear of exchangeable sintered carbide inserts during the cutting proces. *Manufacturing Technology, 14*(4), 590–596. ISSN 1213-2489.
3. Flenger, P., et al. (2019). Processing a measured vibroacoustic signal for rock type recognition in rotary drilling technology. *Measurement: Journal of the International Measurement Confederation, 134*, 451–467. ISSN: 0263-2241.
4. Pitel, J., Matiskova, D., & Marasova, D. (2019). A new approach to evaluation of the material cutting using the artificial neural networks. *TEM Journal - Technology Education Management Informatics, 8*(2), 325–332 ISSN: 2217-8309.
5. Peterka, J., et al. (2020). Modification of cutting tools by drag finishing. *MM Science Journal, 2020*, 3822–3825. ISSN: 1803-1269.
6. Monka, P. P., et al. (2020). Optimal cutting parameter specification of newly designed milling tools based on the frequency monitoring. *International Journal of Advanced Manufacturing Technology*. https://doi.org/10.1007/s00170-020-06169. ISSN: 0268-3768.
7. Monková, K., et al. (2020). Durability and Tool wear investigation of HSSE-PM milling cutters within long-term tests. In *Engineering failure analysis* (Vol. 108) ISSN: 1350-6307.
8. Michalík, P., Zajac, J., Duplák, J., & Pivovarník, A. (2012). CAM software products for creation of programs for CNC machining. In *Lecture notes in electrical engineering, 2011,international conference on future communication, computing, control and management: ICF4C 2011: Pucket: 16–17 December, 2011* (Vol. 141, no. 1, pp. 421–425) ISSN: 1876-1100 ISBN: 978-364227310-0.
9. Suh, Y. (2019). Laser sensors for displacement, distance and position. *Sensors, 19*, 1924. https://doi.org/10.3390/s19081924
10. Kishore, P., Dinkar, D., Shankar, M., Srimannarayana, K., Pachava, V. R., Sengupta, D., & Parne, S. (2011). Non-contact vibration sensor using bifurcated bundle glass fiber for real time monitoring. *Proceedings of SPIE*, 8311. https://doi.org/10.1117/12.904566
11. Blateyron, F. (2011). Chromatic confocal microscopy. In R. Leach (Ed.), *Optical measurement of surface topography*. Springer. https://doi.org/10.1007/978-3-642-12012-1_5

12. Hošek, J., & Linduška, P. (2021). Simple modification of a commercial laser triangulation sensor for distance measurement of slot and bore side surfaces. *Sensors, 21*, 6911. https://doi.org/10.3390/s21206911
13. Knapčikova, L., & Peraković, D. (2022). Innovative approaches in technology challenges in the context of industry 4.0. *Wireless Network, 28*, 427–429. https://doi.org/10.1007/s11276-021-02863-0
14. Knapcikova, L., & Peraković, D. (2021). Challenges of industrial engineering, management and ICT. *Wireless Network, 27*, 1557–1559. https://doi.org/10.1007/s11276-021-02598-y
15. Zhou, Q., Chen, R., Huang, B., Liu, C., Yu, J., & Yu, X. (2019). An automatic surface defect inspection system for automobiles using machine vision methods. *Sensors, 19*, 644.
16. Zbontar, K., Podobnik, B., Povše, F., & Mihelj, M. (2013). On-machine laser triangulation sensor for precise surface displacement measurement of various material types. *Proceedings of SPIE, 8839*. https://doi.org/10.1117/12.2022838
17. Yu, Q., Wang, C., Zhang, Y., Hu, S., Liu, T., Cheng, F., Wang, Y., Lin, T., & Xi, L. (2022). Error analysis and correction of thickness measurement for transparent specimens based on chromatic confocal microscopy with inclined illumination. *Photonics, 9*, 155. https://doi.org/10.3390/photonics9030155
18. Mcgrath, M., Scanaill, N., & Cliodhna. (2013). Sensing and Sensor Fundamentals. In *Sensor Technologies*. Apress. https://doi.org/10.1007/978-1-4302-6014-1_2
19. Zou, X. C., Zhao, X. S., Li, G., Li, Z. Q., & Sun, T. (2017). Non-contact on-machine measurement using a chromatic confocal probe for an ultra-precision turning machine. *International Journal of Advanced Manufacturing Technology, 90*, 2163.
20. Jordan, H.-J., Wegner, M., & Tiziani, H. (1998). Highly accurate non-contact characterization of engineering surfaces using confocal microscopy. *Measurement Science and Technology, 9*(7), 1142–1151.
21. Fu, S., Cheng, F., Tjahjowidodo, T., Zhou, Y., & Butler, D. (2018). A non-contact measuring system for in-situ surface characterization based on laser confocal microscopy. *Sensors, 18*, 2657. https://doi.org/10.3390/s18082657
22. Zhou, Q., Chen, R., Huang, B., Liu, C., Yu, J., & Yu, X. (2019). An automatic surface defect inspection system for automobiles using machine vision methods. *Sensors, 19*(3), 644. https://doi.org/10.3390/s19030644
23. Cheng, F., Fu, S., & Chen, Z. (2020). Surface texture measurement on complex geometry using dual-scan positioning strategy. *Applied Sciences, 10*, 8418. https://doi.org/10.3390/app10238418

Use of IIoT Equipment for Error Messaging in Industrial Production

Stanislav Miščík, Jozef Dobránsky, and Martin Pollák

1 Introduction

Madakam and Uchiya say that the Industrial Internet of Things (IIoT) finds its main application especially in the manufacturing industry. For most industrial sectors, it is very attractive due to better operational efficiency of the production process. It typically features a mechanism for identifying intelligent objects using built-in technologies. It is interesting for its intelligent automation and continuous monitoring capabilities. Its biggest advantage is that it reduces labor interventions in high-risk industrial environments. The best places and activities for IIoT deployment include production halls, material handling, assembly lines, production processes, finalization of goods, and other incoming and outgoing logistics tasks. The basis for the phenomenon of IIoT growth are technologies behind the Internet of Things (IoT), which currently ensure efficient work in many spheres, industrial, as well as commercial and social [1].

Pizon et al. state that the use of IIoT is extremely important due to advancing informatization of hardware resources. This leads to the development of a virtualized model of autonomous production management in real time [2].

Dhungana et al. submit that an important pre-condition for determining whether a certain product can be produced in a given production facility is an accurate assessment of which production lines and/or machines are able to carry out the

S. Miščík · J. Dobránsky (✉)
Technical University in Košice, Faculty of Manufacturing Technologies with a seat in Prešov, Department of Automotive and Manufacturing Technologies, Prešov, Slovakia
e-mail: jozef.dobransky@tuke.sk

M. Pollák
Technical University in Košice, Faculty of Manufacturing Technologies with a seat in Prešov, Department of Computer Aided Manufacturing Technologies, Prešov, Slovakia

L. Knapčíková, D. Peraković (eds.), *7th EAI International Conference on Management of Manufacturing Systems*, EAI/Springer Innovations in Communication and Computing, https://doi.org/10.1007/978-3-031-22719-6_7

necessary production steps. Not only static information about the capabilities of the machines, but also the condition of the machines and tools are important for the present analysis. Due to variations in the capabilities of the machine as its wear progresses, it is also necessary to continuously monitor and analyze the machine's condition [3].

Another article by Qi et al. deals with cloud solutions in IIoT. They state that the basic IIoT technology is the radio frequency identification (RFID) technology. This allows industrial enterprises to identify items and anchor IoT data to those items in time series. They can share Internet of Things data through a cloud service. This enables the exchange of information and supports critical decisions in production operations. On the other hand, storing IoT data in the cloud requires a data access control mechanism to protect sensitive business issues. The disadvantage of using traditional cryptographic access control schemes for IoT data in a time series is that they face serious problems with efficiency and key leakage [4].

Beier et al. submit that industrial production plays an important role in achieving the goals of green economy and sustainable development. Therefore, the nascent transformation of industrial production as a result of digitization into the so-called Industrial Internet of Things (IIoT) is very interesting in terms of sustainable development [5].

Senthilkumar and Rajesh state that Industrial Internet of Things (IIoT) integrated with cloud computing resources offers effective advances in industrial automation with open connectivity and emergent computing. With such advances, ancillary services are evolving to expand the automation process in the manufacturing industry. The integration and transformation of the cloud and the IIoT environment, on the other hand, represents a serious limitation to reliability and computing power. It is, therefore, necessary to develop a framework capable of providing a bridge between rapid acquisition of inputs from the IIoT device and rapid processing or storage of cloud data [6].

Ren et al. emphasize the need to pay more attention to cloud edge computing in providing large IIoT data. Interaction latency needs to be reduced to meet real-time requirements. Artificial intelligence (AI) technologies are expected to significantly change the way big IIoT data is processed [7].

Li et al. deal with the design and verification of a secure communication scheme for a system of intelligent production line of the industrial Internet of Things with multipath redundancy and cooperation. The phenomenon of diversified terminal structures is common in the Industrial Internet of Things (IIoT) and the ways of transferring information between terminals are more complicated. Wireless network transmission and the security of wireless data communication between various intelligent devices are major challenges facing IIoT [8].

2 Deployment of IIoT Devices into the Existing Cloud Infrastructure of the Enterprise

In this chapter, we will describe how to get such messaging using IIoT device, which will be implemented in the existing cloud infrastructure of the enterprise, which hosts the industrial system platform Wonderware System Platform 2017 for designing, managing, and collecting data of production technology units of the enterprise [9].

Production management and control takes place in the control rooms of individual production departments, using HMI/SCADA operator stations connected to the enterprise cloud infrastructure [10]. Operator stations are operated by individual shifts operators. The screens display warnings and alarms in the operator station alarm line (see Fig. 1).

Due to inattention and routine, important error messages may be overlooked. To avoid this, messages will be sent to multiple operators, as well as other authorized persons, via real-time email and SMS notifications. The topological diagram (see Fig. 2) shows the principle of their operation [11, 12].

SMS Notification

The IIoT device for SMS messaging consists of a GSM modem with a SIM card of a mobile operator and a serial converter to Ethernet (TCP Server). The devices are interconnected by a serial cable. The TCP server is connected to the enterprise network with an Ethernet cable. On the virtual server hosting the GSM-

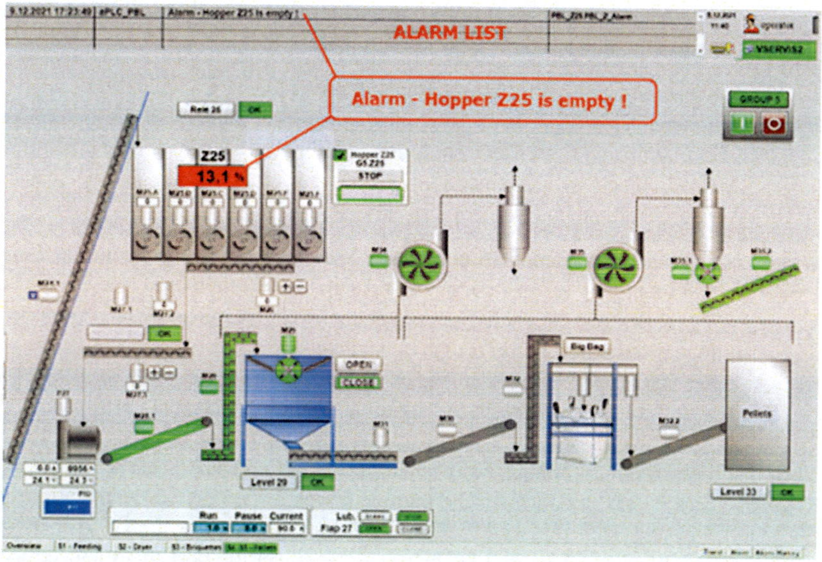

Fig. 1 Example of alarm messages on the operator station

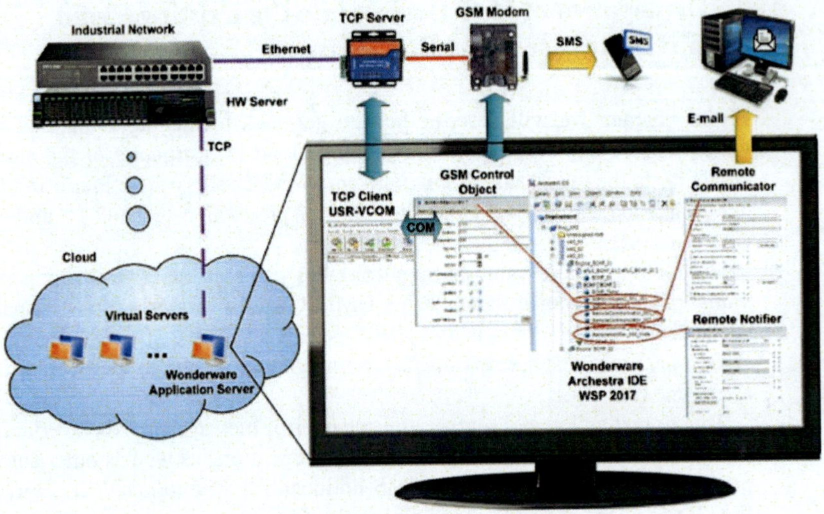

Fig. 2 Topological scheme of error messages operation

Control Object application (by Klinkmann), a USR-VCOM virtual serial port server application must be installed to ensure communication with the TCP server and the GSM modem as a serial device.

GSM Modem Cinterion EGX81W RS232

The GSM modem with a SIM card will serve as a terminal for sending SMS messages to authorized employees' cell phones in real time [13].

Ethernet to USR-TCP232-306 Serial Converter

The converter will be used to connect the GSM modem to the enterprise cloud infrastructure with a virtual computer running on an HW server hosting a GSM-Control Object (by Klinkmann – Wonderware Finland), i.e., an object installed in Wonderware System Platform 2017 [14, 15].

It provides communication between WSP 2017 and IIoT devices and also takes care of managing alarms, contacts, and sending text messages (see Fig. 3) [16].

The converter is configurable via the web interface. After connecting to a computer ethernet cable or via a switch, enter the default IP address of the device into the web browser, which can be found on the bottom of the converter (192.168.0.7), together with the login data (admin, admin) [17, 18].

After logging in to the converter's administration interface, set the new IP address, the mask, the gateway, and possibly also the DNS server.

In the next window, set the parameters of serial port communication (see Fig. 4), i.e., the baud rate = 115,200 bps (must also support the GSM modem), data size = 8, parity = none, stop bits = 1. Next, set the port to 23 and the converter work mode to the TCP server.

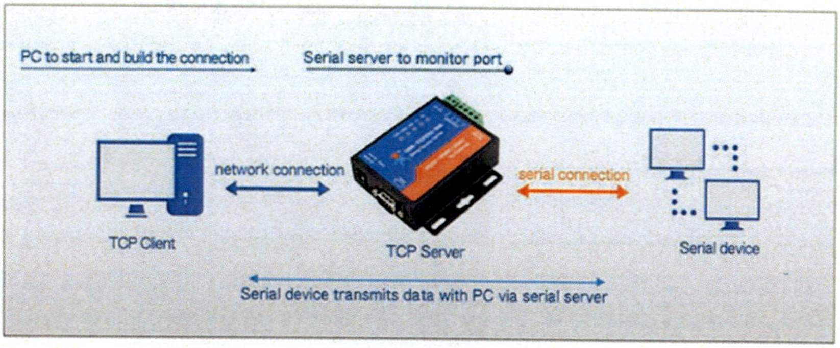

Fig. 3 Communication between WSP 2017 and IIoT devices [18]

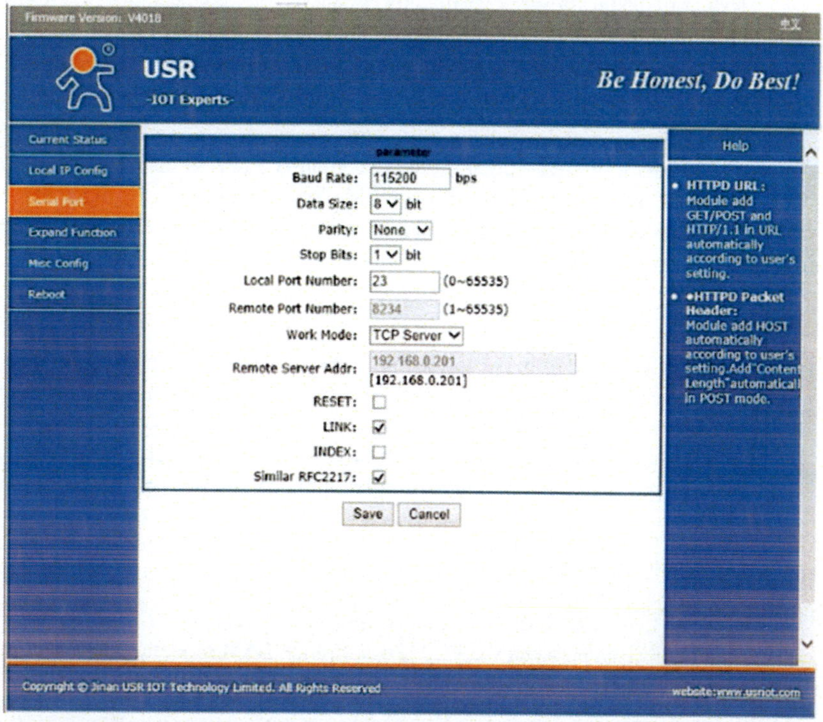

Fig. 4 Setting serial port communication parameters

Confirm the converter configuration with the Save button. We now have a converter ready to be used in the network. Connect the converter to the GSM modem with a SIM card using the serial cable included in the package. Connect the

Fig. 5 Interconnecting the individual devices

devices to the power supply using power adapters and connect them to the enterprise network using an ethernet cable to the enterprise switch (see Fig. 5).

USR-VCOM Virtual Serial Port Server

Install the USR-VCOM application, which can be downloaded from the converter manufacturer's website (https://www.pusr.com/products/ethernet-to-serial-converters-usr-tcp232-306.html) to the computer or, in our case, to the virtual server where the application engine of the GSM-Control Object (by Klinkmann) will run. The application will be used to create a virtual COM port and communicate with the converter (TCP server) [19].

Having installed the application and running the same, search for our serial converter to ethernet (TCP server) in the network (see Fig. 6) by pressing the Smart VCOM button in the top menu. In the Smart VCOM pop-up window, check our converter found in the network and press the Finish button to complete the installation and configuration.

From now on, we have access to our GSM modem on the COM2 virtual serial port. The last step before SMS notifications can be received is installation and configuration of the GSM-Control Object in the WSP2017.

GSM-Control Object

The GSM-Control Object (by Klinkmann - Wonderware Finland) is a paid version of the object installed in the Wonderware Application Server (Aveva) which ensures communication between the WSP 2017 and the GSM modem. The object manages alarms, contacts, and sends out text messages.

The GSM-Control Object is created using the Archestra Object Toolkit. It is used for 2-way communication in automation and other applications using standard cell phones and the GSM network. Based on standard cellular phones and other GSM devices, GSM-Control Object offers a cheap and easy alternative to creating wireless control and monitoring applications. The implementation of SMS (Short Message Service) technology in the GSM network ensures reliable transmission even in applications that are more error sensitive.

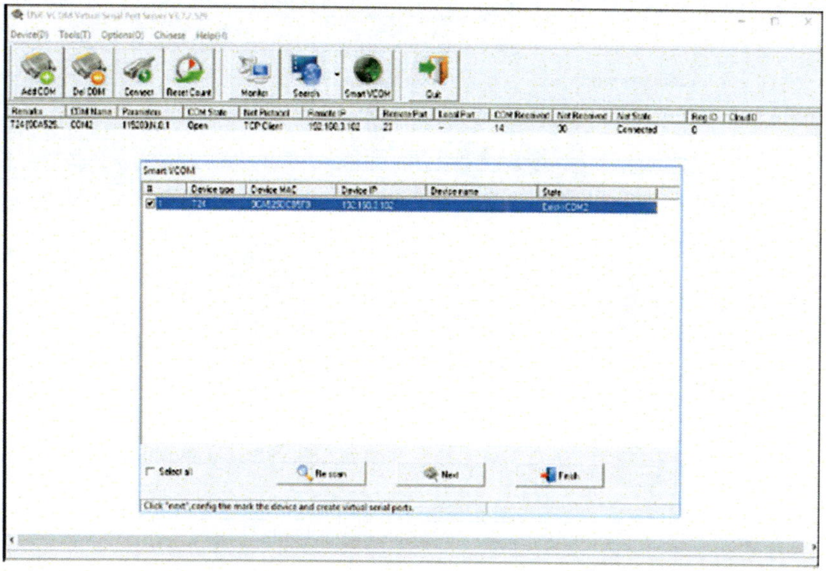

Fig. 6 Serial converter configuration in the network

The main task of the GSM-Control Object is to function as a gateway between the GSM environment on the one hand (connection with the GSM environment through sending/receiving SMS messages) and the Wonderware System Platform environment on the other (see Fig. 7). The GSM-Control Object is installed, configured, deployed, and running in the environment of the Wonderware Application Server, which allows other Wondeware System Platform software to access SMS data via the GSM-Control Object attributes. Data transmission is based on two-way communication of the GSM SMS messages, where a GSM modem (any GSM modem supporting (SMS) AT service commands) is connected to a standard (RS-232) serial port of the computer (or to a USB port via a USB to a serial port adapter) and a GSM cellular phone (remote GSM phone) is used by remote operators in production. Communication between two GSM modems is also supported, i.e., remote GSM modems can be used instead of a remote GSM phone. In principle, data exchange via the GSM-Control object can be initiated from both the GSM environment and the Wondeware System Platform:

1. From the GSM environment – by sending a text message to the GSM-Control Object, where the received message is checked and processed according to the current GSM-Control Object configuration.
2. From the Wondeware System Platform – when an alarm or event occurs associated with a GSM-Control Object alarm attribute [19].

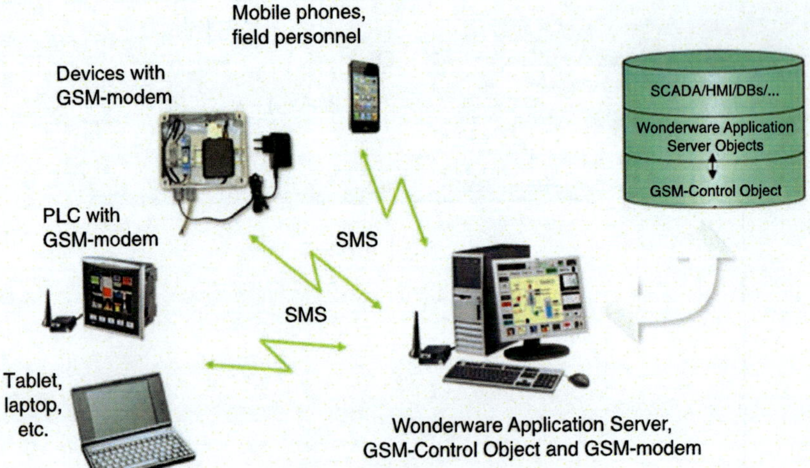

Fig. 7 GSM-Control Object as a communication gateway [13]

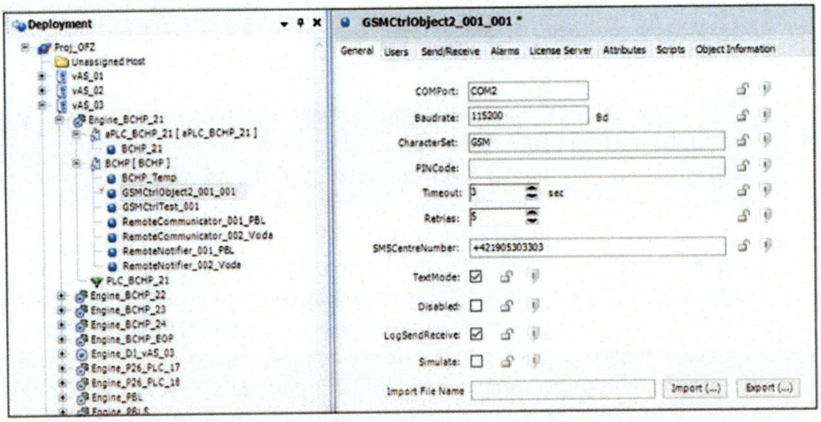

Fig. 8 General tab of the GSM-Control Object

The installation procedure can be found in the GSM-Control Object user manual. After installation, configure the object instance by clicking on GSMCtrlObject2_001_001 in the Wonderware Application Server.

In the General tab (see Fig. 8), set COMPort: COM2 (serial port of the GSM modem that has been connected to the virtual serial port COM2.

The baud rate is set similarly to the TCP server at 115,200 Bd. For CharacterSet, choose GSM. If there is a PIN with the SIM card, the CharacterSet will be set to the PINCode: (e.g., 0000), if no PIN is used or it has been switched off in the SIM, leave the field blank. In SMSCentreNumber, enter the SMS center number of your

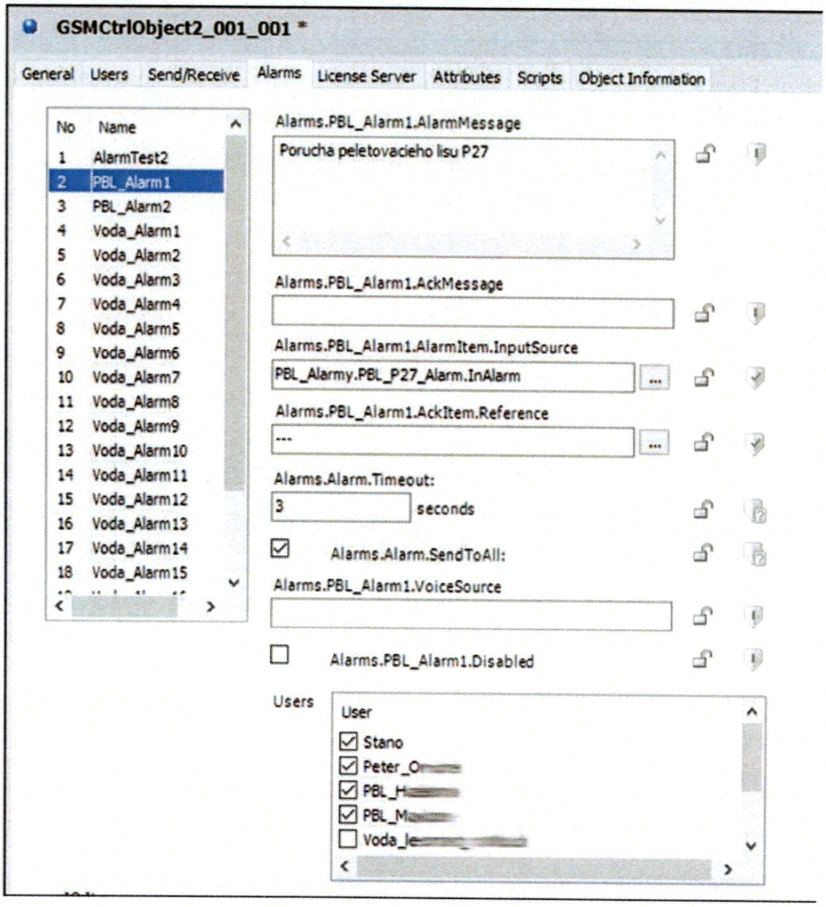

Fig. 9 Alarms tab to the GSM-Control Object

SIM card operator in the international format. Tick the TextMode for messages in text format (AT + CMGF = 1) and LogSendReceive for logging [20, 21].

On the Alarms tab (see Fig. 9) define alarms to be notified via SMS. Set the text for each alarm in AlarmMessage. Select the respective alarm from the Wonderware application server via the InputSource item. Under the Timeout heading, select delayed SMS alarm dispatch to individual recipients according to priority. If we check the SendToAll item, an SMS will be sent instantaneously to all the recipients selected in the last item Users (the priority of the recipients does not matter).

Under the Users tab (see Fig. 10), define the names of the recipients of text messages with telephone numbers in the international format in the PhoneNumber item. In the priority item, set the priority for sending SMS to individual recipients

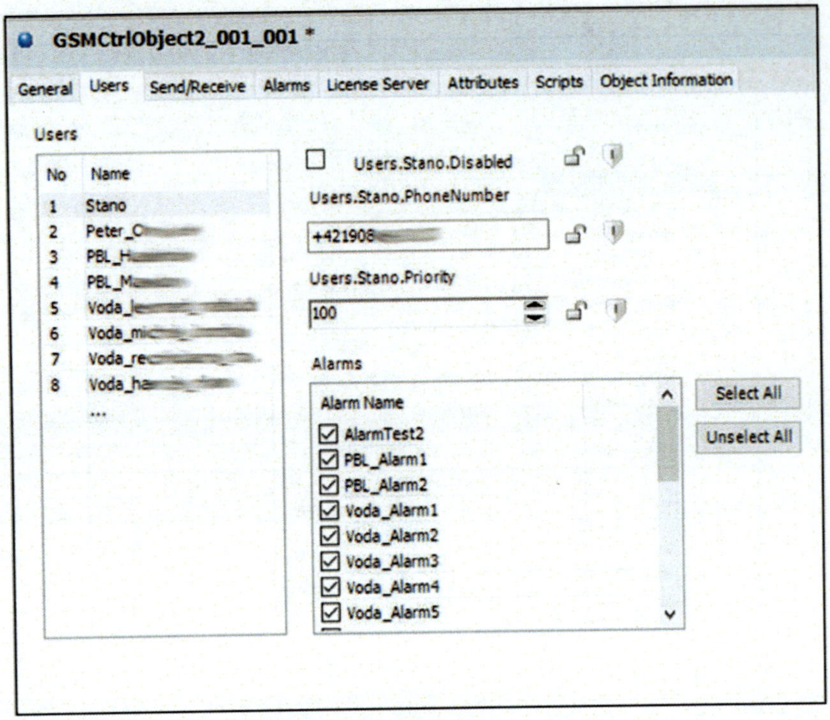

Fig. 10 Users tab of the GSM-Control Object

(lower number = higher priority). Check Alarm Name for individual recipients, i.e., alarms that we want to send the recipients as a text message.

Save GSMCtrlObject2_001_001 and execute Deploy in the application server. The GSM-Control object is in place, and it should start communicating with the GSM modem and the SIM card. For testing, define a test alarm that can be simulated [22].

Email Notification

For email notification, the Remote Response Object needs to be installed in the WSP2017 Application Server (see Fig. 11). Registered users can download it from: https://www.aveva.com/en/products/remote-response-objects/ [23].

Upon installing (or importing) the Remote Response toolkit, the $ Remote Communi-cator and $ RemoteNotifier templates show up in the Application Server, from which the RemoteCommunicator_001_PBL and RemoteNotifier_001_PBL instances can be derived for the production department we want to notify by email.

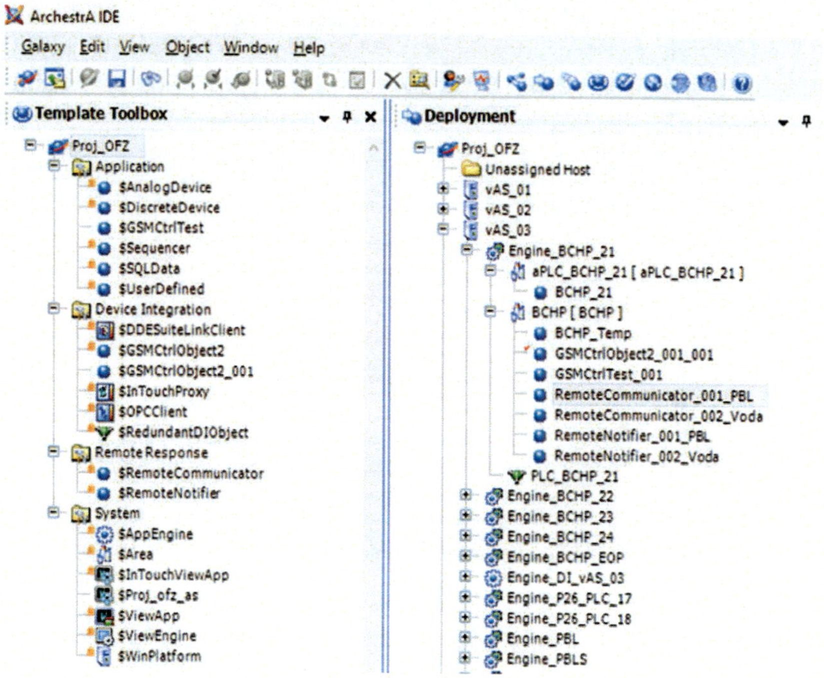

Fig. 11 Application server (WSP2017) in ArchestrA IDE

Remote Communicator

The Remote Communicator object serves as a communication object between the Wonderware System Platform and the email server with defined recipients. After opening the created instance of the RemoteCommunicator_001_PBL object under the General tab (see Fig. 12), the email address (+ password) from which emails will be dispatched needs to be defined. An SMTP server for outgoing mail and a POP3 server for incoming mail needs to be set up with the relevant ports. We can also check the options that suit us. Having filled in everything necessary in a correct way, test the connection with the email server using the Test Connection buttons.

Go to the Contacts and Scheduling tab (see Fig. 13) where it is necessary to fill in the contact details for the email notification, namely the first and the last name, email address and, to enable the recipient's notification, check Always Notify. Save the object and Deploy it in the Application Server.

Remote Notifier

The Remote Notifier object is used to define alarms from the Wonderware System Platform that will be sent to the recipients listed in the Remote Communicator Object. After opening the instance of the RemoteNotifier_001_PBL object under

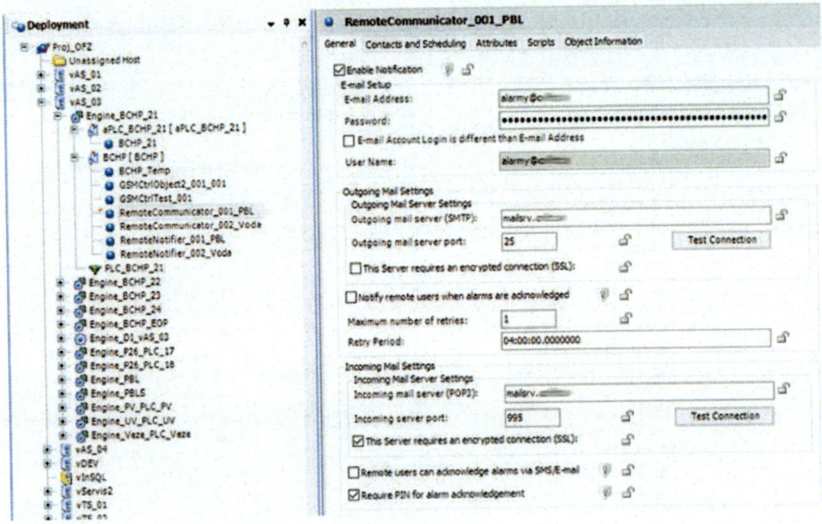

Fig. 12 General tab of the Remote Communicator

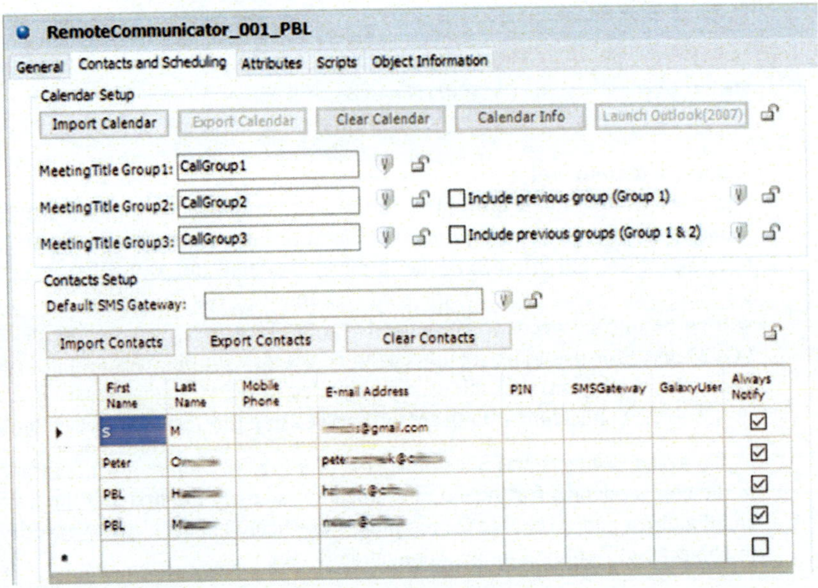

Fig. 13 Contacts and Scheduling tab in Remote Communicator

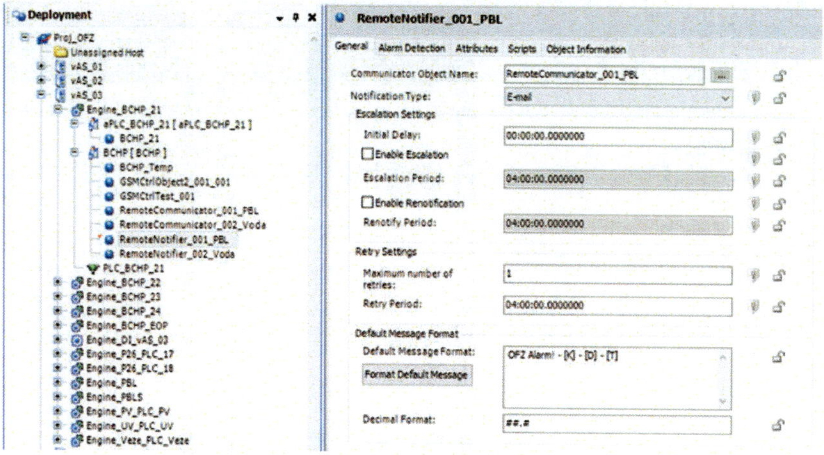

Fig. 14 General tab of the Remote Notifier

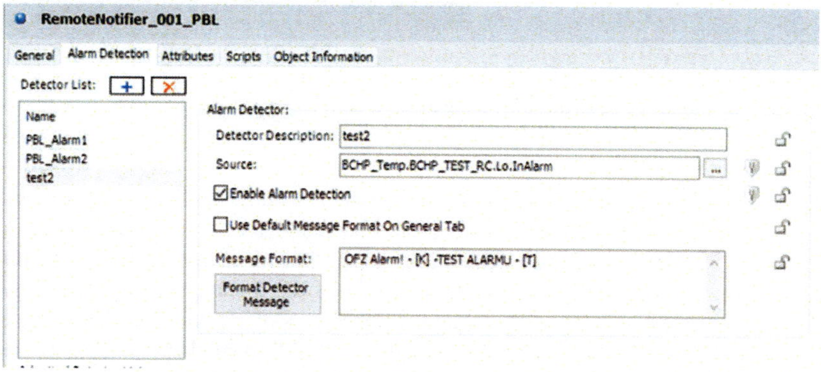

Fig. 15 Alarm Detection tab of Remote Notifier

the General tab (see Fig. 14), select: RemoteCommunicator_001_PBL as the Communicator Object name.

Define individual alarms under the Alarm Detection tab (see Fig. 15). Use the Detector List: + button to create a new alarm and enter its name.

In the Detector Description item, enter the alarm description. Select the specific signal (alarm) that needs to be notified from the WSP 2017 via the Source item. To enable alarm notification, check Enable Alarm Detection. Check Use Default Message Format On General Tab to enable the uniform email text template defined in the General tab. However, we chose a different text for each alarm separately in the form of: OFZ Alarm! - [K] - Alarm Text - [T]. (Individual automatic markers

Fig. 16 Example of a text message and email notification on a smartphone

such as [T] - alarm time, etc. can be found in the object's user manual, as well as explanations of individual items. Save the object and Deploy it in the Application Server.

This completes the basic configuration for sending SMS and emails. Testing may now start, or the test alarm can be introduced into live operation and notifications can be awaited. Check the results of your efforts on the Smartphone (SMS and email) (see Fig. 16) or on a computer (email). The Logger (Log View) in WSP 2017 under the ArchestrA System Management Console (SMC) is also a useful testing tool.

In addition, text messaging can also be done through the set of Remote Response tools from Aveva (formerly Wonderware) directly in the Application Server without any hardware, but due to unsuitability or cost of services offered (SMS gateway) by the mobile operators in our country, the presented HW solution with a GSM modem and a prepaid SIM card with endless SMS was chosen instead.

The mobile operators Mail2SMS service, which forwards emails to SMS messages in real time, can be also used without any hardware. Just enter the recipient's second contact email address in the form 421XXXXXXXXX_XXXX@mail2sms.xxxxxx.xx in Remote Communicator under the Contacts and Scheduling tab. When an email is sent to this email address, which contains a telephone number, an SMS message is automatically forwarded via the SMS gateway. However, the given service is unfortunately a paid service according to the number of text messages sent. In the long run, and with the number of alarms notified by SMS, the HW solution with a GSM modem is a better choice.

3 Conclusions

By integrating the proposed solution (using IIoT devices) to the existing cloud infrastructure of the enterprise, we can eliminate or minimize operator errors in production management and get an overview of the actual status and idle time of production departments from the notified alarms and events.

The main benefits of the solution:

- The awareness of the operator, managers, and management about the condition and failure rate of individual production sections in the factory has improved.
- Based on the information gathered, measures for the most frequent faults were implemented to correct them and modernize faulty equipment.
- Communication between individual workers in different shifts has been introduced. Employees began to communicate with each other, either by phone or email, about the failures and downtime.
- Communication between staff has reduced the time required to rectify faults and downtime.
- The notification helped to improve the efficiency and utilization of production lines (increased labor productivity). Prior to the notification, the night shift mostly ended earlier, despite the higher production capacity of the lines. Following the notification of managers and management about the operation of the line, the production capacity of individual sections increased.
- Workers achieved greater responsibility for the operation of production sections.
- In conclusion, just to add that such applied notification using IIoT equipment significantly helped the factory to modernize production, higher work efficiency, and thus better competitiveness and flexibility, because today it is increasingly difficult to adapt to more demanding market requirements.

Acknowledgments This chapter has been elaborated in the framework of the project KEGA no. 063TUKE-4/2021 and KEGA no. 038TUKE-4/2021.

References

1. Madakam, S., & Uchiya, T. (2019). Industrial Internet of Things (IIoT): Principles, processes and protocols. *Internet of Things in the industrial sector: Security and device connectivity, smart environments, and Industry 4.0*, 35–53. https://doi.org/10.1007/978-3-030-24892-5_2
2. Pizon, J., Klosowski, G., & Lipski, J. (2019). Key role and potential of Industrial Internet of Things (IIoT) in modern production monitoring applications. *III international conference of computational methods in engineering science (CMES 18). MATEC web of conferences*, 252. https://doi.org/10.1051/matecconf/201925209003
3. Dhungana, D., Haselbock, A., Meixner, S., Schall, D., Schmid, J., Trabesinger, S., & Wallner, S. (2021). Multi-factory production planning using edge computing and IIoT platforms. *Journal of Systems and Softvware, 182*. https://doi.org/10.1016/j.jss.2021.111083

4. Qi, S. Y., Lu, Y. S., Wei, W., & Chen, X. F. (2021). Efficient data access control with fine-grained data protection in cloud-assisted IIoT. *IEEE Internet of Things Journal, 8*, 2886–2899. https://doi.org/10.1109/JIOT.2020.3020979
5. Beier, G., Niehoff, S., & Xue, B. (2018). More sustainability in industry through Industrial Internet of Things? *Applied Sciences-Basel, 8*. https://doi.org/10.3390/app8020219
6. Senthilkumar, P., & Rajesh. (2021). Design of a model based engineering deep learning scheduler in cloud computing environment using Industrial Internet of Things (IIOT). *Journal of Ambient Intelligence and Humanized Computing*. https://doi.org/10.1007/s12652-020-02862-7
7. Ren, L., Liu, Y. X., Wang, X. K., Lu, J. H., & Deen, M. J. (2021). Cloud-edge-based lightweight temporal convolutional networks for remaining useful life prediction in IIoT. *IEEE Internet of Things Journal, 8*, 12578–12587. https://doi.org/10.1109/JIOT.2020.3008170
8. Li, M. S., Yin, Z. Y., Ma, Y., Wang, C. X., Chai, A. Y., & Lian, M. J. (2021). Design and verification of secure communication scheme for industrial IoT intelligent production line system with multi-path redundancy and collaboration. *Neural Computing & Applications*. https://doi.org/10.1007/s00521-021-05990-z
9. Vinoth, R., Deborah, L. J., Vijayakumar, P., & Kumar, N. (2021). Secure multifactor authenticated key agreement scheme for industrial IoT. *IEEE Internet of Things Journal, 8*, 3801–3811. https://doi.org/10.1109/JIOT.2020.3024703
10. Chen, B. T., Wan, J. F., Lan, Y. T., Imran, M., Li, D., & Guizani, N. (2019). Improving cognitive ability of edge intelligent IIoT through machine learning. *IEEE Network, 33*, 61–67. https://doi.org/10.1109/MNET.001.1800505
11. https://www.thalesgroup.com/en/markets/digital-identity-and-security/iot/iot-connectivity/products/iot-products/egx81 (16.12.2021).
12. Duplakova, D., Knapcikova, L., Hatala, M., & Szilagyi, E. (2016). Mathematical modeling of temperature characteristics of RFID tags with their subsequent application in engineering production. *TEM JOURNAL-Technology Education Management Informatics, 5*, 411–416. https://doi.org/10.18421/TEM54-01
13. Cepova, L., Kovacikova, A., Cep, R., Klaput, P., & Mizera, O. (2018). Measurement system analyses - gauge repeatability and reproducibility methods. *Measurement Science Review, 18*, 20–27. https://doi.org/10.1515/msr-2018-0004
14. https://www.pusr.com/products/ethernet-to-serial-converters-usr-tcp232-306.htm/ (17.12.2021).
15. Monkova, K., Monka, P., & Hloch, S. (2014). Inverse processing of undefined complex shape parts from structural high alloyed tool steel. *Advances in Mechanical Engineering*. https://doi.org/10.1155/2014/478748
16. Pandova, I., Panda, A., Valicek, J., Harnicarova, M., Kusnerova, M., & Palkova, Z. (2018). Use of sorption of copper cations by Clinoptilolite for wastewater treatment. *International Journal of Environmental Research and Public Health, 15*. https://doi.org/10.3390/ijerph15071364
17. https://www.aveva.com/en/products/sms-gateway-gsm/ (18.12.2021).
18. Pitel, J., Mizakova, J., & Hosovsky, A. (2013). Biomass combustion control and stabilization using low-cost sensors. *Advances in Mechanical Engineering*. https://doi.org/10.1155/2013/685157
19. Novotny, L., Beres, M., de Abreu, M., Zajac, J., & Bleck, W. (2019). Thermal analysis and phase transformation behaviour during additive manufacturing of Ti-6Al-4V alloy. *Materials Science and Technology, 35*, 846–855. https://doi.org/10.1080/02670836.2019.1593669
20. https://www.klinkmann.com/software-for-wireless-automation/ (21.12.2021).
21. https://www.aveva.com/en/products/remote-response-objects/ (21.12.2021).

Application of Numerical Methods in Determining the Regression Model of Body Cooling

Dušan Knežo

1 Introduction

Numerical mathematics provides effective tools applicable in many areas. The paper [1] presents the application of numerical methods in the statistical evaluation of calibration experiments of pressure conductive rubber. In the article [2], numerical methods are used in determining the mathematical model of the amount of harmful substances in the soil. In the work [3], the application of numerical methods in determining the mathematical model of solution diffusion through the membrane is presented. In the work [4], numerical methods are used in the calculation of critical values of several probability distributions. In the work [5], the application of numerical methods in the inverse transformation of the normal probability distribution is presented. This work is a continuation of the mentioned works, and it shows the application of numerical methods in determining the mathematical model of body cooling.

2 Theoretical Model of Body Cooling

Many articles are devoted to the issue of heat transfer and body cooling. From scientific and historical point of view, the overview of the issue can be found in the works [6] and [7]. If there is an exchange of heat between the body and the medium in which the body is placed, then

D. Knežo (✉)
Faculty of Manufacturing Technologies with a seat in Prešov, Technical University of Košice, Prešov, Slovak Republic
e-mail: dusan.knezo@tuke.sk

L. Knapčíková, D. Peraković (eds.), *7th EAI International Conference on Management of Manufacturing Systems*, EAI/Springer Innovations in Communication and Computing, https://doi.org/10.1007/978-3-031-22719-6_8

$$\frac{dQ}{dt} = -h \cdot S \cdot (T - T_{env}) \tag{1}$$

applies. Based on this, then

$$\rho \cdot c \cdot V \frac{dT}{dt} = -h \cdot S \cdot (T - T_{env}) \tag{2}$$

or

$$\frac{dT}{dt} = -\frac{h \cdot S}{\rho \cdot c \cdot V} \cdot (T - T_{env})\,. \tag{3}$$

Suppose that in a closed and thermally insulated environment from its external environment filled with a homogeneous medium M a homogeneous body B is located. Let at time $t = 0$ the medium temperature is equal to T_{0E} and the body temperature is equal to T_{0B}, where $T_{0E} < T_{0B}$. Assume that the medium M and the body B do not contain internal heat sources and that there are no temperature gradients in them and that there are no structural or phase changes in the medium or in the body in the temperature range $[T_{0E}, T_{0B}]$ [8]. In that case, under Newton's law, it holds

$$\frac{dT_B}{dt} = -\frac{h \cdot S}{\rho_B \cdot c_B \cdot V_B} \cdot (T_B - T_E)\,, \tag{4}$$

$$\frac{dT_E}{dt} = -\frac{h \cdot S}{\rho_E \cdot c_E \cdot V_E} \cdot (T_E - T_B)\,, \tag{5}$$

where:

- $T_B = T_B(t)$ is the temperature of the body at time t.
- $T_E = T_E(t)$ is the temperature of the medium at time t.
- h is the total heat transfer coefficient between the body and the medium.
- S is the total area over which the transition between the body and the medium takes place.
- ρ_B is the density of the body B.
- ρ_E is the density of the medium M.
- c_B is the specific heat capacity of the body B.
- c_E is the specific heat capacity of the medium M.
- V_B is the volume of the body B.
- V_E is the volume of the medium M.

From Eqs. (4) and (5), we have a system of two first-order linear differential equations

$$\frac{dT_B}{dt} = -\alpha_B T_B + \alpha_B T_E$$
$$\frac{dT_E}{dt} = \alpha_E T_E - \alpha_E T_B, \tag{6}$$

where

$$\alpha_B = \frac{h \cdot S}{\rho \cdot c_B \cdot V_B}, \tag{7}$$

$$\alpha_E = \frac{h \cdot S}{\rho_E \cdot c_E \cdot V_E}. \tag{8}$$

The solution of the system (6)) that satisfies the Cauchy conditions

$$T_B(0) = T_{0B}, \tag{9}$$

$$T_E(0) = T_{0E}, \tag{10}$$

is

$$T_B(t) = \frac{T_{0B} \cdot \alpha_E + T_{0E} \cdot \alpha_B}{\alpha_B + \alpha_E} + \alpha_B \cdot \frac{T_{0B} + T_{0E}}{\alpha_B + \alpha_E} \cdot e^{-(\alpha_B + \alpha_E) \cdot t}, \tag{11}$$

$$T_E(t) = \frac{T_{0B} \cdot \alpha_E + T_{0E} \cdot \alpha_B}{\alpha_B + \alpha_E} - \alpha_E \cdot \frac{T_{0B} + T_{0E}}{\alpha_B + \alpha_E} \cdot e^{-(\alpha_B + \alpha_E) \cdot t}. \tag{12}$$

It is obvious that after a theoretically infinite time, the temperatures will equalize, and the temperature of the medium M and also the temperature of the body B will be

$$T_\infty = \frac{T_{0B} \cdot \alpha_E + T_{0E} \cdot \alpha_B}{\alpha_B + \alpha_E}. \tag{13}$$

The relations (11) and (12) show that the theoretical dependence of the body temperature $T_B(t)$ and also the medium temperature $T_E(t)$ from time has the form

$$T(t) = a + b \cdot e^{-k \cdot t}, \tag{14}$$

where a, b, and k are the constants, $k > 0$.

3 Regression Model

Assume that we have corresponding experimental data on time t and temperature T, as shown in Table 1, where at least four values of t are different from each other, and we look for a regression function of the type

$$T(t) = a + b \cdot e^{-k \cdot t}, \tag{15}$$

where a, b, and k are the constants.

3.1 Linearizable Model

If we know all the parameters in the relations (7) and (8), then we calculate the steady-state temperature T_∞ from the relation (13). If we use the approximation $a = T_\infty$, then we know the regression function given by the relation (15) by logarithmization and transformation $z = \ln(T - T_\infty)$ linearize to form

$$z = b^* - k \cdot t, \tag{16}$$

where $b^* = \ln b$.

3.2 Nonlinearizable Model

If we cannot determine the steady temperature approximation T_∞ with sufficient accuracy, then in terms of the least squares method we look for the coefficients a, b, and k such that the value

$$S(a, b, k) = \sum_{i=1}^{n} (T(t_i) - T_i)^2 = \sum_{i=1}^{n} \left(a + b \cdot e^{-k \cdot t_i} - T_i\right)^2 \tag{17}$$

is minimal. If we put the partial derivatives of the function S according to the individual variables equal to zero, we get a system of three nonlinear equations

Table 1 Experimental data

t	t_1	t_2	...	t_{n-1}	t_n
T	T_1	T_2	...	T_{n-1}	T_n

$$
\begin{aligned}
n \cdot a + b \cdot \sum_{i=1}^{n} e^{-kt_i} - \sum_{i=1}^{n} T_i &= 0, \\
a \cdot \sum_{i=1}^{n} e^{-kt_i} + b \cdot \sum_{i=1}^{n} e^{-2kt_i} - \sum_{i=1}^{n} T_i \cdot e^{-kt_i} &= 0, \\
a \cdot \sum_{i=1}^{n} t_i \cdot e^{-kt_i} + b \cdot \sum_{i=1}^{n} t_i \cdot e^{-2kt_i} - \sum_{i=1}^{n} t_i \cdot T_i \cdot e^{-kt_i} &= 0.
\end{aligned} \tag{18}
$$

We can solve the system (18) by some numerical method [9], but in this case there may be complications with estimating the position of the solution and also with the convergence of the method. The following procedure may be considered more appropriate.

Assuming that

$$
n \cdot \sum_{i=1}^{n} e^{-2kt_i} - \left(\sum_{i=1}^{n} e^{-kt_i} \right)^2 \neq 0 \tag{19}
$$

holds for each $k > 0$, from the first two equations of the system (18), we express the unknown a and b as follows:

$$
a = \frac{\sum_{i=1}^{n} T_i - b \cdot \sum_{i=1}^{n} e^{-kt_i}}{n}, \tag{20}
$$

$$
b = \frac{n \cdot \sum_{i=1}^{n} T_i \cdot e^{-kt_i} - \sum_{i=1}^{n} T_i \cdot \sum_{i=1}^{n} e^{-kt_i}}{n \cdot \sum_{i=1}^{n} e^{-2kt_i} - \left(\sum_{i=1}^{n} e^{-kt_i} \right)^2}. \tag{21}
$$

By excluding unknown a a b from the third equation of the system (18), we get a nonlinear equation

$$
F(k) = 0, \tag{22}
$$

where

$$
F(k) = n \cdot \sum_{i=1}^{n} t_i T_i e^{-kt_i} \cdot \sum_{i=1}^{n} e^{-2kt_i} -
$$

$$
- \sum_{i=1}^{n} t_i T_i e^{-kt_i} \cdot \left(\sum_{i=1}^{n} e^{-kt_i} \right)^2 - \sum_{i=1}^{n} T_i \cdot \sum_{i=1}^{n} t_i e^{-kt_i} \cdot \sum_{i=1}^{n} e^{-2kt_i} -
$$

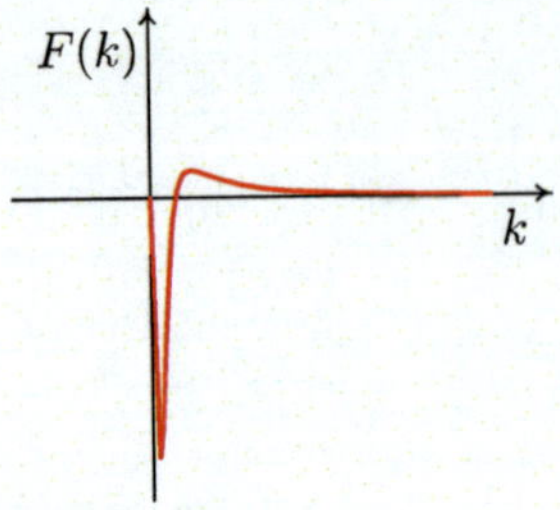

Fig. 1 Graph of the function F for test data

$$-n \cdot \sum_{i=1}^{n} t_i e^{-2kt_i} \cdot \sum_{i=1}^{n} t_i T_i + \sum_{i=1}^{n} t_i e^{-2kt_i} \cdot \sum_{i=1}^{n} e^{-kt_i} \cdot \sum_{i=1}^{n} T_i +$$

$$+ \sum_{i=1}^{n} e^{-kt_i} \cdot \sum_{i=1}^{n} t_i e^{-kt_i} \cdot \sum_{i=1}^{n} T_i e^{-kt_i}. \tag{23}$$

Obviously, $F(0) = 0$ and $\lim_{k\to\infty} F(k) = 0$. The function F is continuous on the interval $[0, \infty)$, and if the system (18) has only one solution, then Eq. (22) has on the interval $(0, \infty)$ unique solution (see Fig. 1).

4 Program for Calculation of Regression Model Parameters

To determine the parameters a, b, and k of the regression function given by the relation (15), a program was written in Free Pascal language, which on the basis of known experimental data and given accuracy $\varepsilon > 0$ of calculation of a parameter k performs the following:

- Specifies the interval $[\alpha, \beta]$, in which the root of Eq. (22) lies.
- Using the bisection method with accuracy ε calculates the root of Eq. (22), i.e., the value of a parameter k.
- Calculates the parameters a and b from the relations (20) and (21) based on the known value of a parameter k.

The program was verified on test data. The graph of the regression function for the test data is shown in Fig. 2.

5 Conclusion

The presented method of application of numerical methods in determining the regression function of the body cooling makes it possible to find the regression

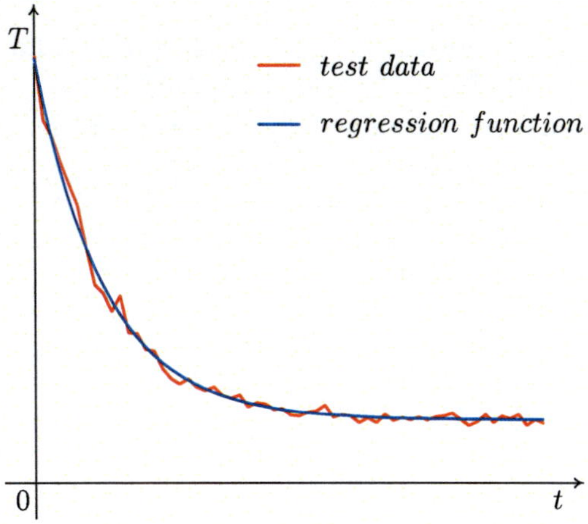

Fig. 2 Regression function for test data

function in cases where the conditions do not allow the temperature to be determined with sufficient accuracy after equalizing the temperature of the body and the medium. Obviously, the method can also be applied in determining arbitrary regression functions of a type $f(x) = a + be^{-kx}$. Last but not least, the method is presented by inspiration how to use numerical methods in determining different types of nonlinear regression functions.

Acknowledgments This work was financed by Scientific Grant Agency of Ministry of Education of Slovak Republic and Slovak Academy of Science—project VEGA 1/0301/20.

References

1. Knežo, D., Šimšík, D., Živčák, J. (1995). Mathematical evaluation of pressure conductive rubber calibration experiments. *Bulletins for Applied Mathematics, 1068/95 [LXXV]*, 43–50.
2. Knežo, D. (2012). Mathematical model of harmful substance amount in soil. *Acta Mechanica Slovaca, 16*(1), 26–29.
3. Knežo, D. (2013). Model of solution diffusions through membranes. *Acta Mechanica Slovaca, 17*(3), 98–101.
4. Knežo, D., Zajac, J., Michalik, P. (2017). Calculation of critical values of several probability distributions using standard numerical methods. In: *SmartCity 360 2016* (pp. 1–7). Gent: EAI.
5. Knežo, D. (2014). Inverse transformation method for normal distribution and the standard numerical methods. *International Journal of Interdisciplinarity in Theory and Practice, 2014*(5), 6–10.
6. Cheng, K. C., Fujii, T. (1998). Isaac Newton and heat transfer. *Heat Transfer Engineering, 19*(4), 9–21.
7. Michael, I. Davidzon, M. I. (2012). Newton's law of cooling and its interpretation. *International Journal of Heat and Mass Transfer, 55*(21–22), 5397–5402.
8. Kopal, I., Harničárová, M., Valíček, J., Kušnerová, M. (2016). Model of dynamic thermal interaction between a solid and a fluid. *Hutnické listy 6/2016, LXIX*, 28–33.
9. Mac Berthouex, P., Brown, L. C. (2002). *Statistics for environmental engineers* (2nd ed.). Boca Raton, Florida: Lewis Publishers.

Part III
Sustainable Industrial Development

Development of Sink-Float Density Separation Process of Mixed Automotive Plastics for Mechanical Recycling

Dorota Czarnecka-Komorowska and Wiktoria Kanciak

1 Introduction

Nowadays, plastics are one of the most important engineering materials commonly used in the production of automotive parts [1]. Plastics are one of the most important construction materials, and the second most frequently used material in car construction, as well as their quantitative share in electrical, electronic, and telecommunication devices is very significant. An increasing number of electrical and electronic devices ensuring the efficient operation of all the systems in a car are installed in new car models [2, 3], which increases the amount of electrical wiring used. In an average passenger car, the mass of cables amounts to about 25 kg [3, 6]. According to the Central Statistical Office, in 2017, there were more than 10.000 tons of cable installations in Poland to be recycled [8]. This is due to the properties they possess, which allow them to meet specific requirements concerning the functions they can perform in the finished product [24]. Cables are recovered mainly through material recycling, sometimes chemical recycling, and in some cases, energy recovery is carried out.

The more frequent use of polymeric materials, for example in electrical devices, makes it possible to reduce weight and dimensions. These materials guarantee good electrical conductivity as well as thermal and electrical insulation or can guarantee a suitable flammability class. Thermoplastics are a group of polymers that are most commonly used in electronics. Examples are polypropylene (PP), acrylonitrile-butadiene-styrene (ABS), polycarbonate (PC), polybutylene terephthalate (PBT), and polyamide (PA). These polymers constitute a group of materials of high quality and thus of higher price compared to so-called mass plastics. This is one of the

D. Czarnecka-Komorowska (✉) · W. Kanciak
Poznan University of Technology, Poznan, Poland
e-mail: Dorota.Czarnecka-Komorowska@put.poznan.pl

L. Knapčíková, D. Peraković (eds.), *7th EAI International Conference on Management of Manufacturing Systems*, EAI/Springer Innovations in Communication and Computing, https://doi.org/10.1007/978-3-031-22719-6_9

main factors encouraging their recycling or reuse. This type of waste is a mixture of different types of polymers, the recovery of which brings economic advantages as well as benefits to the environment [2–6].

Due to their versatility of applications and variety of properties, their global production has not been decreasing over the years. The increase in the consumption of plastics results in an increase in the amount of these materials in the total waste. To reduce the negative impact of polymer waste on the environment, plastics are recycled, which allows, for example, their reuse in the manufacture of new products [7–9]. The recovery of materials from waste is an attractive process for manufacturers due to the reduction of material costs, as well as the reduction of environmental impacts [8, 10].

Nowadays, the aim is to recycle as much as possible of the waste from electronics, as well as automotive, for recovery and reuse. The wide range and variety of polymer grades used, as well as the use of additives that affect the properties or processing characteristics, are the main problems during the recycling of this type of waste [11–13]. The growing awareness of society and continuous improvement of polymeric waste management methods result in an increased frequency of subjecting these materials to recovery and reuse processes, instead of landfilling. The application of the above-mentioned processes results in ecological benefits, due to, e.g., reduced consumption of natural resources, but also brings economic advantages [2, 3, 14, 15].

Nowadays, recycling is one of the most dynamically developing areas in the plastics industry, because the appropriate segregation and separation of mixed materials are the basic steps in the preparation of polymeric waste for this process [16, 17]. There are many scientific papers in the literature on the process of segregation and separation of mixed plastic waste [7, 18].

As an example, a method of separating various plastics from mixed post-consumer waste using three-stage flotation combined with selective flotation can be given. The tested mixture included among others: high-density polyethylene (HDPE), poly(vinyl chloride) (PVC), polypropylene (PP), polystyrene (PS), polyethylene terephthalate (PET), and acrylonitrile-butadiene-styrene copolymers (ABS). To separate the mixed plastics, first, a three-stage flotation in water was carried out. As a result, plastics were divided into two fractions – light and heavy. Then, to separate the obtained fractions into individual polymers, a selective flotation was carried out, in which different types and concentrations of wetting agents affecting the efficiency of the process were used. The most effective results of selective separation of PET and PVC were obtained using 500 mg/l of calcium lignosulfonate as a wetting agent, whereas for ABS and PS separation the most effective process was carried out during flotation with 200 mg/l of calcium chloride as a wetting agent [19].

Various concentrations and combinations of wetting agents and inorganic agents can be used to separate materials contained in cables and wires that are a combination of different types of metals and plastics. Based on the results obtained, the most favorable compositions and concentrations of the additives used were determined. The best results for the separation of PVC from other materials were obtained using the first method 10 mg/l of tannic acid and 50 mg/l of calcium lignosulfonate and in the second method using flotation with the addition of calcium chloride solution. During the separation of aluminum from copper, flotation in the presence of 48 wt.% calcium chloride solution proved to be the most effective calcium chloride solution [8].

Separation of polymers can also be done based on their density difference. The process carried out by the float-sink method is most effective using sodium and calcium lignosulfonate as wetting agents. Another method for the separation of mixed polymer waste by density is the process of using cyclone separators. Cyclones generate centrifugal forces that accelerate the rate of separation of particles of different densities. Two varieties of these devices can be distinguished: cylindrical and conical. Test results show that these devices can provide separation with a significant increase in production efficiency and a much wider range of particle sizes down to 0.5 to 120 mm [20].

Another method allowing the separation of mixed plastics is the electrostatic method. Regulski et al. proposed a concept of a stand for separation by the electrostatic method of mixed plastics used in the automotive sector. Electrostatic separation allows to separate efficiently metals from plastics but also to separate polymers into single fractions. The automated station for separation of mixed plastics, designed by the authors, was also equipped with an original computer vision system adapted to analyze the influence of parameters on the efficiency of the separation process. Investigations were carried out to determine the influence of the high voltage value and the influence of the electrode position on the efficiency of the separation process to determine the parameters that would allow obtaining the highest efficiency of the separation process of the mixed materials. The separation process with the described method allows for the effective separation of single polymers from mixed waste containing poly(methyl methacrylate) PMMA and polystyrene (PS) [21].

To evaluate the efficiency of electrostatic separation of mixed polymer plastics, a dedicated vision system can be used. The study was conducted on a mixture containing polystyrene (PS) and poly(methyl methacrylate) (PMMA), differing in color. The operation of the vision system is enabled by the use of a camera and a proprietary computer system. Based on the results obtained, it was concluded that the vision systems can be successfully applied to evaluate the efficiency of the electrostatic separation process. The solution proposed by the authors can be used not only to evaluate the results of electrostatic separation but can also be applied to other methods used for the separation of plastics. To increase the accuracy of the described solution, various lighting systems can be applied, as well as the method can be used for the separation of a greater number of materials differing in color [22].

The study aimed to develop a method for the separation of mixed polymer waste from the automotive industry by flotation, together with an evaluation of the mechanical properties of the separated plastics.

2 Experimental and Methods

2.1 *Materials*

The mixed polymeric waste from the mechanical cable recycling process was the subject of the study (Fig. 1). The mixture consists of the following polymer types: polypropylene (PP), polyamide 6 (PA 6), polyamide 66 (PA 66), poly(vinyl chloride) (PVC), and poly(butylene terephthalate) (PBT).

2.2 *Processing and Sample Preparation*

2.2.1 A Sink-Float Density Separation Technique

A 3 kg mixture of polymeric waste was introduced to the separation process by flotation. The separation process was performed using a device constructed to separate plastics by the flotation method as shown in Fig. 2. The experiment was carried out at the Plastics Division of Poznan University of Technology. During the separation process, the temperature of the liquid (flotation medium) was 20 °C, the screw rotational speed was 50 Nm, and the blade rotational speed was 40 Nm.

An experiment was conducted using two process steps:

- Stage I, in which the waste was subjected to sedimentation, in two stages: in water and water with the addition of NaCl
- Stage II, where an additional agent: a surfactant was introduced between stages: in water and water with NaCl

Due to unsatisfactory results, the type I process was discarded and in the further part of this work, the materials formed in the type II process with the addition of surfactant were studied. To interpret graphically the mass balance, a Sankey flow

Fig. 1 Mixture of polymeric cable wastes introduced into the flotation process

Fig. 2 View of the flotation and sedimentation device for separation process: (**a**) flotation tank, (**b**) separated light fraction, and (**c**) separated heavy fraction

diagram was made as shown in Fig. 3. It illustrates the efficiency of the carried-out process of type II separation of mixed plastics.

The first stage of the separation process was carried out in the water. The heavy fraction separated constituted the vast majority in terms of mass, because out of 3 kg of mixed polymeric waste subjected to the flotation process, 2.87 kg of the heavy fraction was obtained, while 0.12 kg of the light fraction was separated. The second stage of the separation process was carried out in the water with the addition of surfactant. The introduction of this surfactant allowed to lower the surface tension of water to change the wetting angle of the separated material, thus enabling its separation. The separated heavy fraction, obtained during this flotation step, also represents a large majority compared to the light fraction. The heavy fraction obtained was 2.95 kg, which represents 98% of the total mixed polymer

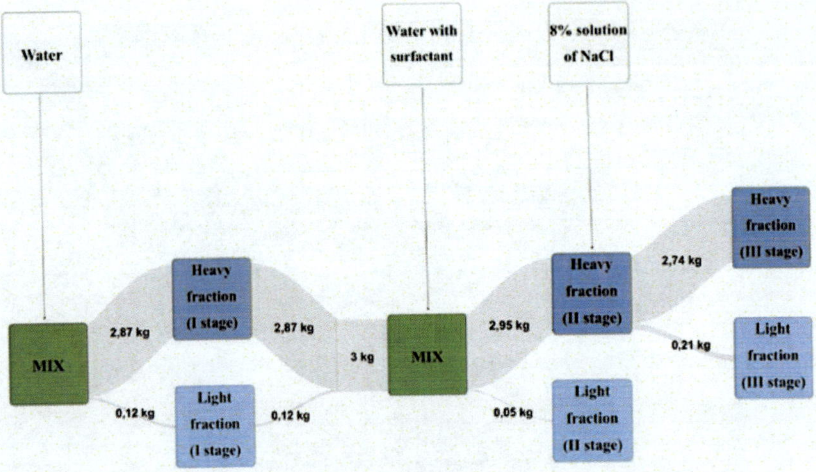

Fig. 3 Sankey Chart illustrating the efficiency of the separation process of mixed plastics as a function of mass, using the flotation method: MIX – a mixture of plastic post-cable waste

waste introduced, while the light fraction, obtained was 0.05 kg, which means that the separated mixed waste mostly has a density greater than 1.00 g/cm^3.

The third stage of the separation process was carried out in a liquid with a density of 1.05 g/cm^{3}which was obtained by introducing sodium chloride (8% in water) into water. In this stage, the separation of the heavy fraction obtained in the previous stage of the process was carried out by rotation, to further separate the polymeric materials contained in this mixture.

As a result of the separation process, a light fraction was separated, containing polymers of density lower than 1.05 g/cm^3 and the heavy fraction containing polymers with a density higher than that of the liquid in the flotation tank. As a result, 2.74 kg of heavy fraction and 0.21 kg of light fraction were obtained.

2.2.2 Sample Preparation for Mechanical Recycling

The polymer waste mixture was separated and divided into light and heavy fractions depending on the density of the liquid in which the flotation process took place. Then, test samples were made by compression molding to form composites (Table 1).

The hot pressing process was carried out using a laboratory press type PLHS-7, using the following parameters, the temperature was 190 °C, and the pressing force was 10 MPa. Plasticizing time was 3 min, forming time was 6 min and cooling time was 24 min. Heat pressing resulted in a 10 × 10 × 2 mm plate from which standardized test shapes were cut. Example microscopic images of the plates of the obtained K2 and K3 composites are illustrated in Fig. 4.

Table 1 Sample codes of the composites considering the density of the liquid in the flotation process

Sample code	The type of sample depends on the density of the liquid in which the flotation process took place
K1	Light fraction, obtained by flotation in a liquid with a density of 1.00 g/cm^3
K2	Heavy fraction, obtained by flotation in a liquid with a density of 1.00 g/cm^3
K3	Heavy fraction, obtained by flotation in a liquid with a density of 1.05 g/cm^3

a)

b)

Fig. 4 View of samples manufactured by hot-pressing: (**a**) K2 composite, (**b**) K3 composite

The K1 composite is obtained from the light fraction by flotation in a liquid with a density of 1.0 g/cm^3. The K2 composite, containing a heavy fraction obtained by flotation in a liquid of density 1.0 g/cm^3, contains the highest amount of polyisoprene, nylon 6, and poly(vinyl chloride). The composite K3, formed from the heavy fraction by flotation in a liquid of density 1.0 g/cm^3, has in its composition the highest amount of poly(vinyl chloride) and poly(butylene terephthalate).

2.2.3 Measurement of the Surface Tension at the Liquid/Gas Interface

The measurement of the surface tension at the liquid/gas interface was performed using the ring method, in which the measuring probe is a horizontally suspended platinum ring. A standard ring with an average ring radius of $R = 9.545$ mm and a wire radius of $r = 0.185$ mm was used for the measurement.

The measurement was performed by immersing the platinum ring in the liquid sample and pulling it out again. The value measured during the measurement is the maximum force required to pull the ring across the surface of the liquid under test. A correction factor is applied to account for the weight of the volume of liquid raised [23].

The interfacial tension at the liquid/gas interface was measured using a K6 force strain gauge made by KRŰSS (KRŰSS GmbH, Germany, Hamburg). The measurement temperature was $T = 22$ °C.

2.2.4 Mechanical Properties

To determine the mechanical properties of the composites obtained, static tensile tests were carried out according to PN EN ISO 527-2. The tests were performed at a tensile speed of 50 mm/min at ambient temperature using a Zwick Roel Z010 universal testing machine.

2.2.5 Microscopic Evaluation of the Surface Structure

To evaluate the surface structure of the composites, microscopic evaluation was performed at 10× and 15× magnification. The fractured structure obtained from the static tensile test was also evaluated at 20× magnification. The study was carried out using a stereoscopic microscope OPTA-TECH SK series with a microscopic camera for image capturing and processing.

3 Results and Discussion

3.1 Measurement Results for the Interfacial Voltage at the Liquid/Gas Interface

Table 2 shows the results of surface tension. Five measurements were taken for each sample and the mean and standard deviation were calculated from the results. For the readings of the interfacial voltage, a correction of the measured value was considered, which was calculated according to the equation of Zuidema and Waters [23] (Fig. 5):

$$F = \sqrt{0.4036 \cdot 10^{-3} \frac{\sigma^*}{D - d} + 0.0128} + 0,725 \quad (1)$$

in which

D – diameter of the measuring ring; $D = 19.09$ mm.

Table 2 Measurement results of surface tension at the liquid/gas interface

Medium	Surface tension with correction [mN/m]
Water	68.755 ± 1.557
Surfactant	58.356 ± 3.094
Sodium chloride solution (8% in water)	79.100 ± 1.432

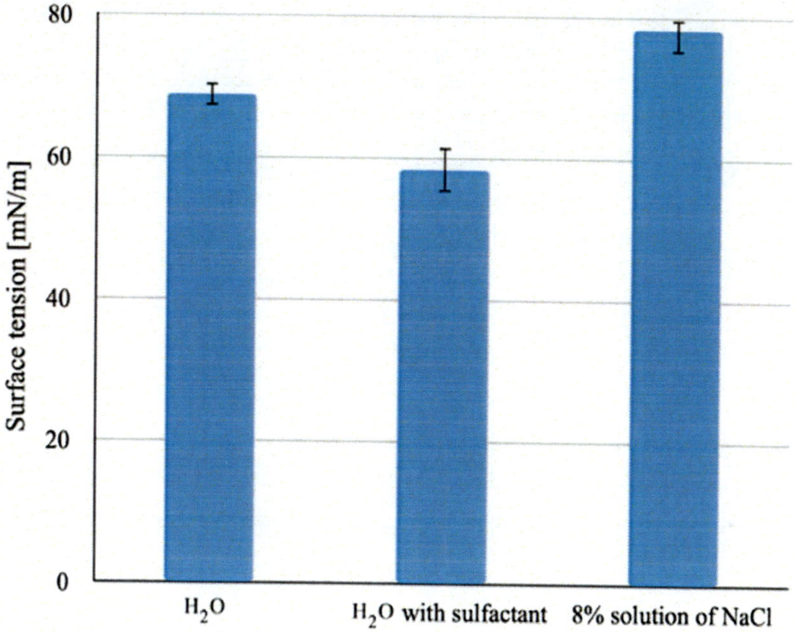

Fig. 5 Comparison of surface tension values at the liquid/gas interface

d – diameter of the test wire; $d = 0.37$ mm
σ – the actual value of the surface tension
σ^*– the measured value of surface tension; $\sigma^* = \sigma \cdot 1.062$ [23]

The surface tension value of water at the liquid/gas interface is higher than that of water with surfactant. The surfactants present in the second sample affect the surface tension value, causing it to decrease. This phenomenon is due to the accumulation of surfactant on the surface of the solution, increasing wettability so that such a solution shows a lower surface tension compared to pure water. The results showed that the highest value of surface tension (79.1 mN/m) is shown by an 8% solution of sodium chloride in water.

3.2 *Results of the Static Tensile Test*

In the tensile test, the stress-strain curves were obtained for investigated K1-K3 blends. Based on them the elastic modulus, short-term strength, elongation at maximum stress, stress at break, and elongation at break were determined. The obtained results are summarized in Table 3. The presented values represent the average of three measurements together with the standard deviation.

Table 3 Static tensile results of K1-K3 composites

Type of composite	Tensile modulus [MPa]	Tensile strength [MPa]	Elongation at max. tensile strength [%]	Tensile at break [MPa]	Elongation at break [%]
K1	113.10 ± 20.00	2.88 ± 0.09	1.45 ± 0.21	1.05 ± 0.15	3.20 ± 0.28
K2	249.50 ± 0.71	2.89 ± 0.43	1.40 ± 0.14	0.58 ± 0.08	4.40 ± 1.41
K3	169.33 ± 11.55	3.13 ± 0.78	2.27 ± 0.76	1.16 ± 0.11	4.80 ± 2.02

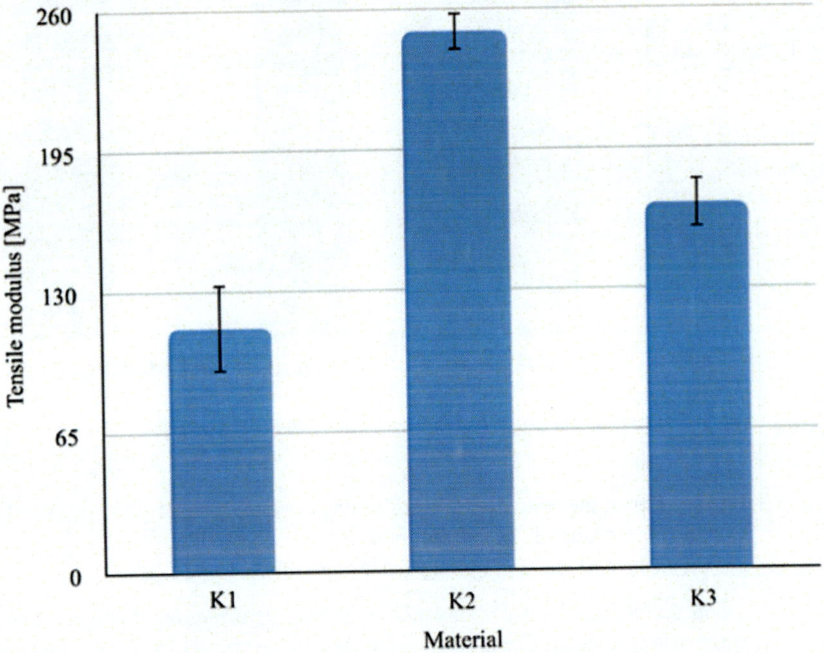

Fig. 6 Comparison of tensile modulus values for K1-K3 composites

Figure 6 shows that the K2-K3 composites obtained from the heavy fraction show higher values of the tensile modulus in comparison with the composite made from the light fraction, which indicates higher stiffness of these materials. The highest tensile modulus has a K2 composite in comparison with the other composites. The lowest value has the K1 composite, which includes the light fraction obtained during the separation in the liquid with a density of 1.0 g/cm^3. The K1 composite also has the highest value of standard deviation in comparison with the other composites, which means that the obtained values of modulus during the test showed the greatest scatter of values.

It can be seen from Fig. 7 that the K1 composite, obtained from the light fraction, and the K2 composite, obtained from the heavy fraction from separation in a liquid

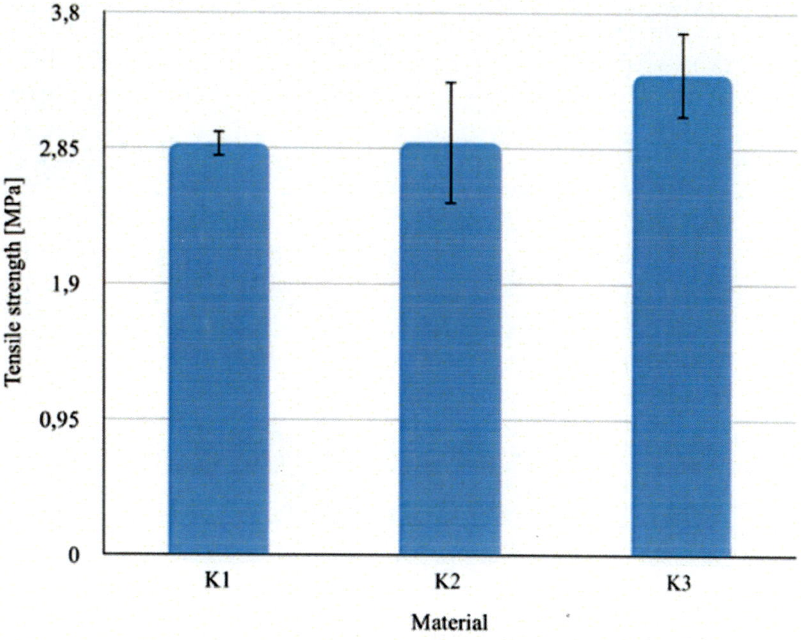

Fig. 7 Comparison of tensile strength values of K1-K3 composites

of density 1.00 g/cm^3 have very similar values of tensile strength. Material K3 is created from a heavy fraction with a density higher than 1.05 g/cm^{3}shows the highest value of tensile strength which is 3.13 MPa (Fig. 7).

Figure 8 shows that the highest value of elongation at maximum tensile strength is shown by composite K3, while it has also the highest standard deviation result, which means that it has the largest spread of measured values. Composite K1 has a similar but slightly higher value of elongation at maximum tensile strength, compared to composite K2 (Fig. 8).

Figure 9 shows that the highest value of tensile at break has the material K3, obtained from the heavy fraction after flotation in a liquid of density 1.05 g/cm^{3}while the lowest result has K2, which also contains the heavy fraction, but is obtained by flotation in water. The highest standard deviation value has the K1 composite, which may be caused by the smallest adhesion between particular components of this composite.

Figure 10 shows that also the highest value of elongation at break has the composite K3, which was formed from the separated heavy fraction with a density greater than 1.05 g/cm^3. The lowest value of the elongation at break is shown by the material K1 containing the light fraction of the polymeric post-cable waste with a

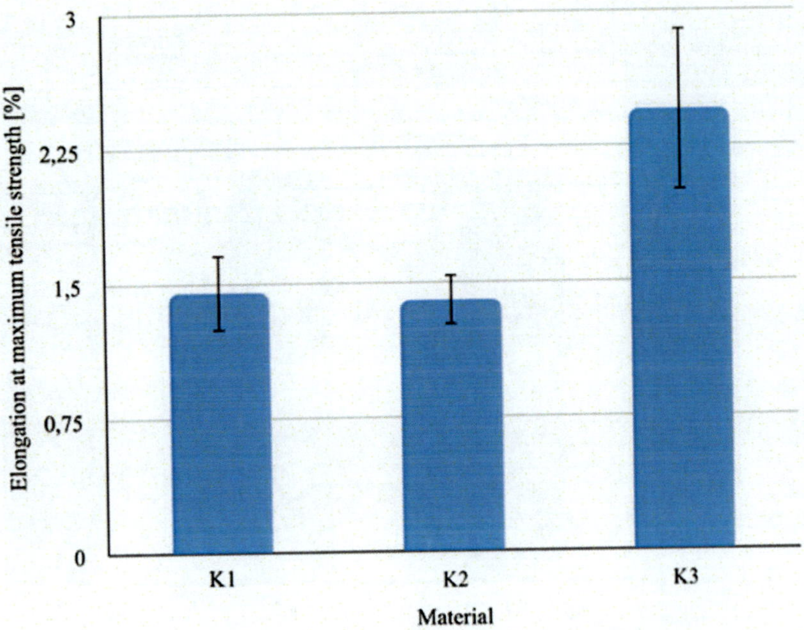

Fig. 8 Comparison of values of elongation at a maximum tensile strength of composites

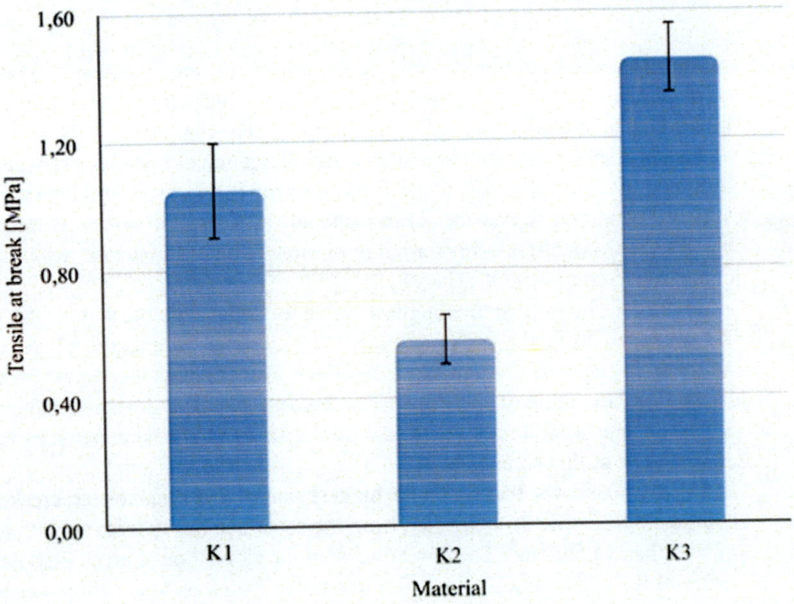

Fig. 9 Comparison of tensile at break values of K1-K3 composites

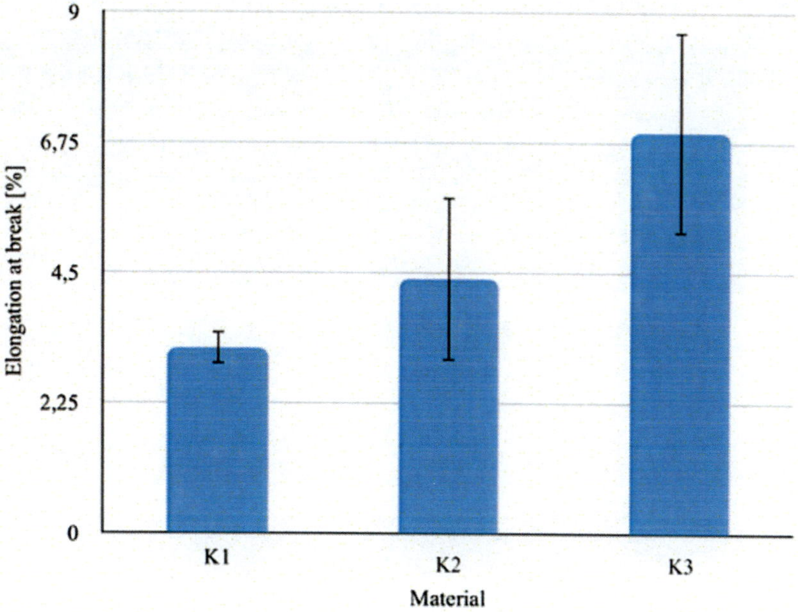

Fig. 10 Comparison of elongation at break values of K1-K3 composites

density less than 1.0 g/cm^3. The composite K2 containing the heavy fraction formed as a result of flotation in water has an elongation at break at the level of 4.40%. Compared to the other composites created, this value is higher in contrast to the K1 composite but much lower in comparison to the K3 material.

3.3 Results of Microscopic Examination of Composites

3.3.1 Results of Tests on the Surface Structure of Composites

The results of microscopic observations of the surface structure of the composites show that the K2-K3 composites produced from the separated heavy fraction are distinguished by a more homogeneous structure, indicating a good interfacial adhesion between the components of the mixture [25, 26] which the authors observed and described in detail in their earlier work [27] (Fig. 11).

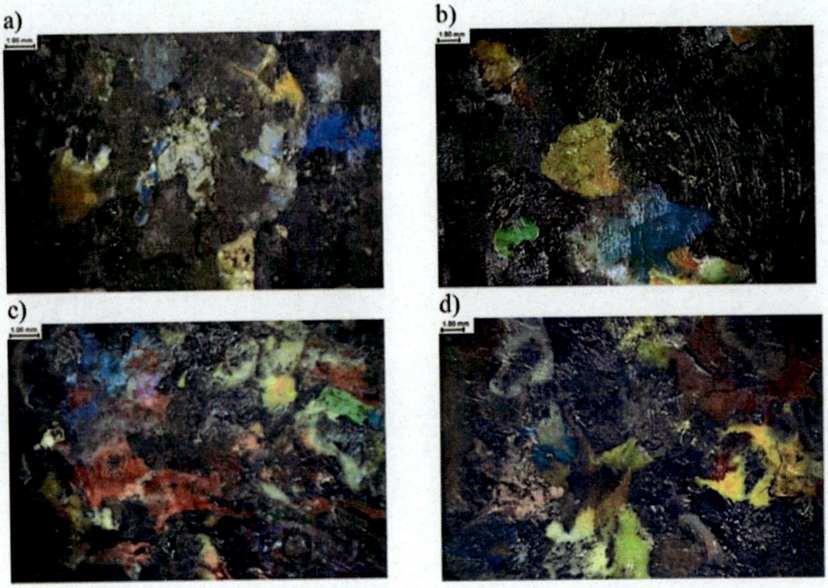

Fig. 11 Microscopic images of the composite surface structure: (a) K2 (10×), (b) K2 (15×), (c) K3 (10×), and (d) K3 (15×)

4 Conclusions

In this work, the efficiency of separation of mixed polymeric waste by flotation method was studied and evaluated. In the next stage of work, the waste became material for the production of new materials. From the mixture of polymers, using various conditions of the flotation process (solutions of various medium concentrations), composites with favorable strength properties and a wide range of applications were produced. The compositions of the individual materials were determined, as well as their structures were tested for strength and evaluated.

The study shows that the addition of surfactant to water, introduced during the separation process by the flotation method, does not change the density of the liquid, but decreases the surface tension at the liquid/gas interface, increasing wettability. The composites made of heavy fractions show higher values of tensile modulus in comparison to the composite made of the separated light fraction. The highest value of tensile modulus is in the K2 composite, which consists of the heavy fraction of mixed polymeric post-cable waste with a density higher than 1.0 g/cm^3. The tests showed that the K3 Composite containing the heavy fraction obtained by flotation in a liquid with a density of 1.05 g/cm^3 is characterized by the highest value of tensile strength. The tests showed that the highest value of elongation at maximum tensile strength and tensile at the break is exhibited by the K3 composite. Higher values of elongation at break have composites formed from the heavy fraction in comparison with the composite obtained from the light fraction. The K3 composite has the highest value of elongation at break in comparison with the other composites created.

From the strength tests carried out and from the microscopic evaluation of the structure of the resulting composites, it is apparent that the composites formed from the heavy fraction, both as a result of flotation in a liquid with a density of 1.0 g/cm^3 and 1.05 g/cm^3, have much more favorable tensile properties in comparison with the composite formed from the separated light fraction. Differences in the structure of the obtained materials can also be observed. The composites formed from the heavy fraction show a much smaller amount of free spaces in the structure in comparison with the composites formed from the separated light fraction. The reason for better tensile properties and more compact structure in the case of K2 and K3 composites formed from the heavy fraction is a greater adhesion between the individual components of the resulting composites.

As a result of the studies carried out, it can be concluded that the separation process using the flotation method in the liquid with the addition of surfactant is the most effective. The optimum management of mixed polymeric waste is obtained for the composite K3 of high stiffness (tensile modulus about 170 MPa) containing separated polymers of density higher than 1.05 g/cm^3.

In conclusion, it is possible to separate mixed polymeric waste by flotation method, using a few-stage separation of only two fractions: light and heavy, depending on the density of the liquid in which the separation process is carried out. To increase the efficiency of the flotation process, surfactants can be introduced into the separation liquid in which the process takes place to increase the wettability of the surface of the separated particles.

References

1. P. Collective edited by M. K. [1], Plastics Recycling in Europe.
2. Kozlowski, M., & Rydarowski, H. (2012). Recycling of polymeric waste from electronics and vehicles. *Wydaw. Nauk. ITB-PIB Radom*.
3. Michalski, J., & Tarkowski, Z. Recycling of electrical and electronic waste.
4. Żenkiewicz, M., & Żuk, T. (2014). Electrostatic separation of poly(vinyl chloride) and poly(ethylene terephthalate) blends. *Przemysl Chemiczny, 93*(1), 54–57. https://doi.org/10.12916/przemchem.2014.54
5. Żenkiewicz, M., Żuk, T., & Blaszkowski, M. (2014). Electrostatic separation of polymer blends with different contents of PVC and PLA, Plastics. *Process Plastics, [R.] 20*(2 (158)) Accessed 13 Dec 2021 [Online]. Available at: http://yadda.icm.edu.pl/baztech/element/bwmeta1.element.baztech-a7ef9642-8bfd-4d31-928a-77a2c3bb4a2a
6. Takoungsakdakun, T., & Pongstabodee, S. (2007). Separation of mixed post-consumer PET-POM-PVC plastic waste using selective flotation. *Separation and Purification Technology, 54*, 248–252. https://doi.org/10.1016/j.seppur.2006.09.011
7. Wang, C., Wang, H., Fu, J., & Liu, Y. (2015). Flotation separation of waste plastics for recycling-A review. *Waste Management, 41*, 28–38.
8. Barbakadze, K., Brostov, V., Granowski, G., Hnatchuk, N., Lohse, S., & Osmanson, A. T. (2018). Separation of metal and plastic wastes from wire and cable manufacturing for effective recycling. *Resources, Conservation and Recycling, 139*, 251–258.

9. Yang, X., Liu, X., Song, L., Lv, C. Y., & Cheng, X. (2017). Study on the separators for plastic wastes processing. *Procedia Engineering, 174*, 497–503. https://doi.org/10.1016/j.proeng.2017.01.172
10. Żenkiewicz, M., Żuk, T., & Królikowski, K. (2012). Methods for sorting waste polymer plastics. *Przetw Plastics, [R.] 18*(6), 692–698.
11. Beđeković, G., Salopek, B., & Sobota, I. (2011). *Electrostatic separation of PET/PVC mixture* (p. 6).
12. Güney, A., Özdilek, C., Kangal, M. O., & Burat, F. (2015). Flotation characterization of PET and PVC in the presence of different plasticizers. *Separation and Purification Technology, 151*, 47–56. https://doi.org/10.1016/j.seppur.2015.07.027
13. Burat, F., Güney, A., & Olgaç Kangal, M. (2009). Selective separation of virgin and post-consumer polymers (PET and PVC) by flotation method. *Waste Management, 29*(6), 1807–1813. https://doi.org/10.1016/j.wasman.2008.12.018
14. Zhang, L., & Xu, Z. (2016). A review of current progress of recycling technologies for metals from waste electrical and electronic equipment. *Journal of Cleaner Production, 127*, 19–36. https://doi.org/10.1016/j.jclepro.2016.04.004
15. Wang, R., & Xu, Z. (2014). Recycling of non-metallic fractions from waste electrical and electronic equipment (WEEE): A review. *Waste Management, 34*(8), 1455–1469. https://doi.org/10.1016/j.wasman.2014.03.004
16. Dodbiba, G., & Fujita, T. (2004). Progress in separating plastic materials for recycling. *Physical Separation in Science and Engineering, 13*(3–4), 165–182. https://doi.org/10.1080/14786470412331326350
17. Al-Salem, S. M., Lettieri, P., & Baeyens, J. (2009). Recycling and recovery routes of plastic solid waste (PSW): A review. *Waste Management, 29*(10), 2625–2643. https://doi.org/10.1016/j.wasman.2009.06.004
18. Lyskawinski, W., et al. (2021). Tribo-electrostatic separation analysis of a beneficial solution in the recycling of mixed poly(ethylene terephthalate) and high-density polyethylene. *Energies, 14*(6), Art. no. 6. https://doi.org/10.3390/en14061755
19. Pongstabodee, S., Kunachitpimol, N., & Damronglerd, S. (2008). Combination of three-stage sink-float method and selective flotation technique for separation of mixed post-consumer plastic waste. *Waste Management, 28*(3), 475–483.
20. Gent, M. R., Menendez, M., Toraño, J., & Diego, I. (2009). Recycling of plastic waste by density separation: Prospects for optimization. *Waste Management & Research, 27*(2), 175–187.
21. Regulski, R., et al. (2021). Automated test bench for research on electrostatic separation in plastic recycling application. *Bulletin of the Polish Academy of Sciences-Technical Sciences, 69*(2), e136719.
22. Rybarczyk, D., et al. (2020). Assessment of the electrostatic separation effectiveness of plastic waste using a vision system. *Sensors, 20*(24), 7201.
23. Principles of measurement with Kruss K6 force tensiometer https://www.kruess.com. Accessed 13/04/2021.
24. https://eldan-recycling.com/sites/default/files/BR_CABLE_1804_EN_low-180409.pdf (dostęp 14.10.2020).
25. Sika, R., et al. (2020). Decision support system in the field of defects assessment in the metal matrix composites castings. *Materials, 13*, 3552. https://doi.org/10.3390/ma13163552
26. Czarnecka-Komorowska, D., et al. (2021). Polyethylene/polyamide blends made of waste with Compatibilizer: Processing, morphology, rheological and thermo-mechanical behavior. *Polymers, 13*(14), 2385. https://doi.org/10.3390/polym13142385
27. Czarnecka-Komorowska, D., et al. (2021). Recycling of plastics from cable waste from automotive industry in Poland as an approach to the circular economy. *Polymers, 13*(21), 3845. https://doi.org/10.3390/polym13213845

Assessment of the Cataphoretic Varnishing Production Process Capability Using the CAQ System of Quality Management

Jozef Dobránsky and Róbert Balint Bali

1 State of the Art

To prevent corrosion of automotive components, cataphoretic coatings have been widely used since the 1970s. Easy automation of the painting process poses a great advantage. This type of painting, i.e., the cataphoresis, does not cause smudges. The thickness of the varnish is evenly distributed over the entire surface, including hard-to-reach places. The surface coverage is very high [1]. Special drainage holes must be introduced into the closed painted structures due to leakage. The biggest disadvantage are UV rays, where graphite-gray coating loses its properties. All that in less than 100 hours [2]. The effectiveness of the coating's anti-corrosion properties depends on the action of anti-corrosion pigments, tightness, and adhesion to the underlying material. The most important and difficult step in the application of the cataphoretic coating is thorough preparation of the surface. A surface that may be contaminated by corrosion, oils, or dust requires multiple operations [3]. Surface in the first group was cleaned chemically. Surface in the second group was sandblasted. Testing also took place in the salt chamber. It has been concluded that the post-surface treatment of cold-rolled steel affects each of the parameters measured. It also affects the properties of the tested specimens, such as roughness, distribution of the coating thickness, friction factor, anti-corrosion resistance, etc. Specimens treated with chemical agents showed a higher quality of surface treatment than specimens that were only sandblasted. The coating after chemical treatment has a homogeneous structure, it does not contain pores or cracks [4]. When examining the adhesion of epoxy cataphoretic coatings to other

J. Dobránsky (✉) · R. B. Bali
Technical university of Košice, Faculty of Manufacturing Technologies with the Seat in Prešov, Prešov, Slovakia
e-mail: jozef.dobransky@tuke.sk

L. Knapčíková, D. Peraković (eds.), *7th EAI International Conference on Management of Manufacturing Systems*, EAI/Springer Innovations in Communication and Computing, https://doi.org/10.1007/978-3-031-22719-6_10

materials, namely on steel modified with Zn-Fe and Zn-Co alloys, it was shown that under dry test conditions, the specimens displayed positive adhesion results. Following their dipping in a 3% NaCl solution, the lowest adhesion values were on the steel substrate. Adhesion to steel modified by the Zn-Co alloy during 24 days of immersion in a 3% NaCl solution saw the smallest change observed from among all the specimens tested [5]. In the research focused on the temperature of the cataphoretic process, tests were carried out in the temperature range of 20–45 °C. It has been found that the process temperature did not affect the adhesion or corrosion resistance of the surface when a salt spray was used. The best mechanical properties and corrosion resistance of the coatings were achieved in the temperature range of 25–30 °C [6]. The most ideal curing temperature at which the low-carbon steel specimen under examination achieved the best surface effects was in the range of 145–195 °C. In fact, the surface formed under this temperature achieves the best anti-corrosion properties. When the temperature was not within the required tolerance limits, these characteristics were significantly reduced. This fact also manifested in tests that led to surface delamination [7]. When examining the effects of graphene oxide in cataphoretic varnishing, it has been concluded that graphene has an effect not conducive to the protective coating layer functionality. This was due to creation of a layer full of defects and, moreover, the layer failed to meet the protective function. A proposal to do double coating was made and implemented at two points or coatings. The first layer was free of graphene particles and the second had graphene added. This procedure has proved to be favorable because the first layer formed a homogeneous structure and the second layer added to the coating's thickness. The result was suitable test effects, not only in the salt chamber, but also in the aggressive environment formed by NaCl, where after seven hundred hours, the specimen showed no signs of damage. Therefore, double application of varnish in the above-mentioned order is classified as a suitable method for the application of coating by cataphoresis, regardless of the reduction in protective properties [8].

2 Varnishing by Cataphoresis

Cataphoresis is a method of electroplating the metals exposed to a large extent to atmospheric influences. It is an electrolyte varnish found on various components. Such coating enhances the components' anti-corrosion properties and is the standard for automotive parts that need complete and optimal anti-corrosion protection [9].

2.1 Coating Options

CATAPHORESIS – the color is positively (+) charged and is called "cationic". The components to be painted must be connected to the negative pole (–) of the electrical circuit and serve as a "cathode".

ANAPHORESIS – the color is negatively charged (−) and is called "anionic". The component to be painted is oriented toward the positive pole (+) of the electrical circuit and serves as an "anode" [10].

2.2 *Cataphoresis Steps (Fig. 1)*

The technological process of cataphoresis is one belonging to the electrophoretic varnishing methods where the desired layer of paint is formed after the phosphating process. This layer serves as anti-corrosion protection of various components. This process is based on the principle of electronic coating. It is, therefore, necessary to use a cationic cooler known as KTL or Cata. Painting takes place in a bath in which the material is immersed and acts as a cathode. Subsequent use of electric current will cause the formation of an electric field. As a result, the polycations start moving toward the cathode. Reaction of electrodes with hydroxide ions (see Fig. 2) guarantees their deposition on the surface of metallic materials. With the required coating getting bigger, the layer, too, creates resistance, which also results in the rate of deposition. Reduction in velocity is noticeable especially in places with small layer thickness – cavities or hard-to-reach places. The entire process guarantees a gradual formation of an evenly distributed coating, which is spread over the entire surface of the material. The final version of the effects occurs only when the desired layer thickness is reached. Subsequently, the deposition process stops. Adherence to technological procedures in the form of constant mixing by ramps located at the bottom of the tank results in the necessary effect of cataphoresis. Once removed from the cataphoretic bath, the components are rinsed by sprinkling. This ensures paint recovery, as it drips from the material on which it is used into the bath. The necessary second rinse is done by immersion to remove excess paint that has not adhered to the component surfaces. The third rinse is repetition of the sprinkling method, which removes the undesirable impurities attaching to the material coming

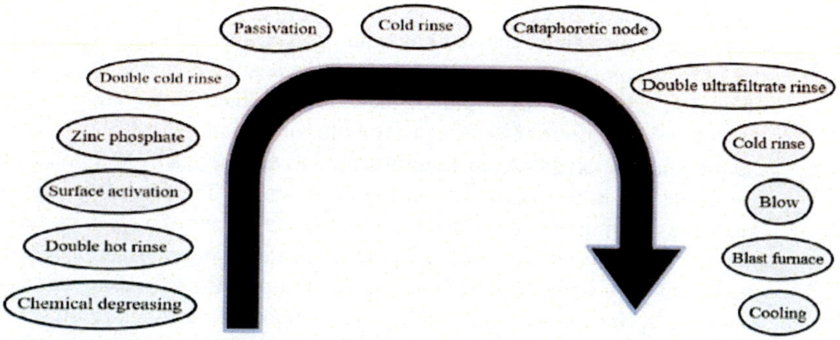

Fig. 1 Schematic representation of the steps of cataphoresis

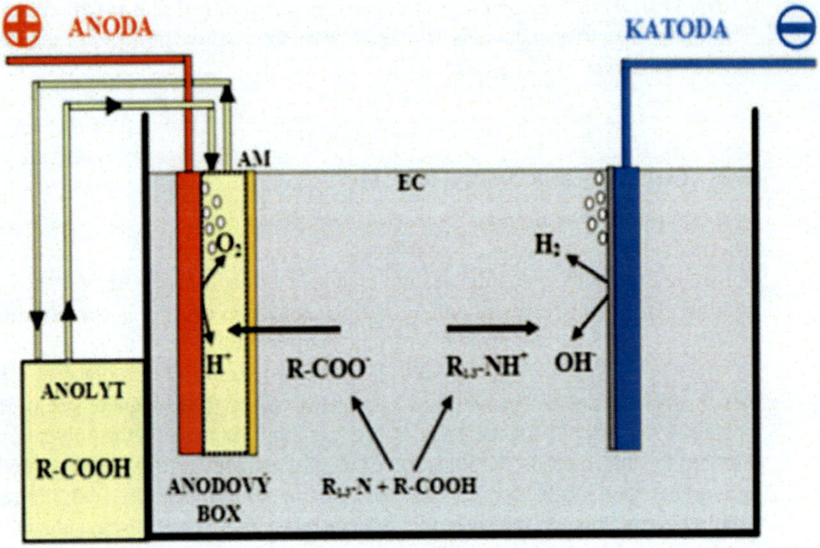

Fig. 2 Principle of cataphoretic deposition of coating materials [11]

from the bath. To ensure the purity of the material, the last rinse, bath, and sprinkling should be done exclusively with demineralized water. The final coating properties are achieved only after polymerization, which takes place during calcination in the temperature range between 155 and 180 °C. This process creates an anti-corrosion layer and a sufficient substrate for subsequent surface treatments [11].

2.3 *Characteristics of Aqueous Solutions Used in Cataphoretic Varnishing*

For cataphoresis paints, aqueous solutions have a limited duration of use, which is one of their typical properties. Properties of the coating material change as a result of the following processes. One of the reasons is that the working solutions age during working operations of the electrolytic bath. Oxidation of film-forming air in the presence of oxygen leads to an increase in the molecular weight of the resin and a decrease in the molecular weight of the binder. The process is supported by continuous stirring of the working solution and the release of oxygen by electrolysis of water. Saponification of ester groups in an alkaline working solution results in the removal of fatty acids. Increasing the acidity number and decreasing the molecular weight of the binder takes place at high speed under elevated temperatures and pH. The hydrolysis of the salt group reduces the solubility of the resin in water during the coagulation of the binder due to a change in the hydrophilic-hydrophobic balance.

Imbalance in the pigment–binder ratio leads to degradation of the coating content and its protective properties. Evaporation of organic solvents causes an increase in the size of the craters and a decrease in the thickness of the coating and reduces the paint-water solution's stability.

The accumulation of foreign electrolytes in the working solution due to the anodic decay of the substrate metal and poor cleaning of the products during surface pretreatment leads to an increase in the electrical conductivity of the working solution. This leads to accumulation of sludge and coagulation of the binder. Accumulation of the neutralizer results in a reduced equilibrium time of the working solution and an improvement of the process parameters, as well as the properties of the coating with a change in the degree of the binder's neutralization. Stability of the solution properties in the bath process is maintained by keeping the operating solution parameters within defined limits and by their periodic adjustment [10].

2.4 *Technological Process of Cataphoretic Varnishing at KTL (Table 1, Figs. 3 and 4)*

Figures 3 and 4 shows the finished components after cataphoretic varnishing. The numbers 1 to 5 indicate the specimens taking sites. 10 components were taken and measured from each site. The measured layer thickness values were recorded in the Palstat software.

Table 1 Technological process of cataphoretic operations

Operation	Position	Technological process	Time/min.	Temp/°C	Voltage/V
010	1.	Hanging up	–	–	–
020	2.	Chemical degreasing – level 1	10:00	70	–
030	3.	Chemical degreasing – level 2	10:00	70	–
040	4.1	Technological rinsing	0:05	–	–
050	4.2	Technological rinsing	0:05	–	–
090	8.	Activation before phosphating	2:00	–	–
100	9.	Phosphating	5:00	50	–
110	10.1	Technological rinsing	0:05	–	–
120	10.2	Technological rinsing	0:05	–	–
130	11.	Rinse with demineralized water	0:01	–	–
140	12.	Cataphoresis painting	4:00	30	230
150	13.1–13.2	Rinse in ultrafiltrate	0:30	–	–
160	14.1–14.3	Burning	40:00	180	–
170	15.1, 15.2	Cooling	0:01	–	–
180	16.	Hanging down	–	–	–

Fig. 3 Comparison of specimens before cataphoretic varnishing on the left and after cataphoretic varnishing on the right

Fig. 4 Specimens taking sites

3 MSA Analysis

Before evaluating the capability of the production process, it was necessary to do the measurement system analysis using MSA. The analysis of the measurement system was evaluated using the Gage R&R repeatability and reproducibility method according to the ARM methodology enshrined in the international IATF 16949 standard.

The thickness of the coating was measured by non-destructive methods, using a digital measuring device called Elcometer 456. The principle lies in measuring 10 specimens by 3 inspectors done three times. The measured values are shown in the table in Fig. 5.

		Sample									
Controller	Measuring	1	2	3	4	5	6	7	8	9	10
- Reference values											
0	0										
- Measured values											
1	1	16.7	22.3	24.1	22.3	21.2	28.9	22.7	21.5	24.2	15.8
	2	16.9	22.8	24.5	22.7	21.6	28.9	23.4	21.9	23	16.4
	3	16.5	22.3	22.4	22.5	20.9	28.5	23.5	22.1	22.6	16.5
2	1	17.5	22.5	22.6	22.4	22.5	28.3	22.8	22.9	22.7	15.5
	2	17.1	22.3	22.8	22.9	22.4	28.5	22.7	22.6	23.3	16.1
	3	17.3	22.4	22.6	22.4	21.8	27.9	22.4	22.8	23.2	15.5
3	1	15.8	23.1	23	22.5	22.5	26.5	22.7	22.7	23.1	15.2
	2	16.2	22.2	22.5	22.6	22.7	27.1	23.1	22.5	22.6	15.8
	3	16.1	23.1	22.4	22.9	22.2	27.3	22.5	22.5	22.7	16.1

Fig. 5 Values measured for the MSA analysis

3.1 *Evaluation*

- The % value of repeatability and reproducibility has reached a GRR of 9.943% (<10%), which lies within acceptable limits considering the type of instrument and application.
- The values of measuring equipment variability (EV) 9.515%, inter-inspector variability (AV) 2.887%, and inter-component variability (PV) 99.504% are at good levels.
- The value of the number of distinguishable categories (ndc) is 14.110. This value speaks of the fact that the measured pieces are produced throughout the entire production process range.

Based on the measured and evaluated values, we can say that the measurement system is competent and can be used to assess the capability of the production process.

3.2 *Graphs and Analyses*

The result of the GRR analysis should also include a graphical analysis. Graphical analysis methods will validate the results of the study and provide further insight into the data. Graphical visualization is an important form of data visualization. The presented graphs demonstrate the effectiveness of the measuring device, not the process of specimens creation. Figure 6 shows a diagram of the mean values achieved by individual inspectors.

The individual measurements are displayed as mean values by inspectors in the mean diagram. The mean diagram provides a better overview of the individual inspectors' consistency. It illustrates the relationships between the individual

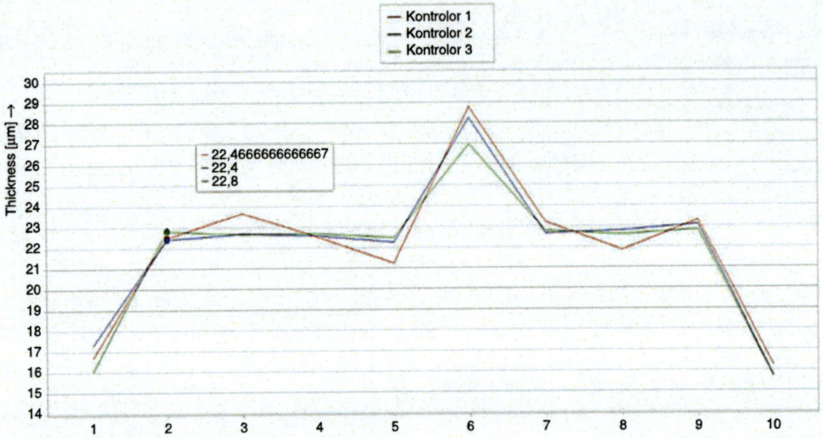

Fig. 6 Diagram of mean values by inspectors

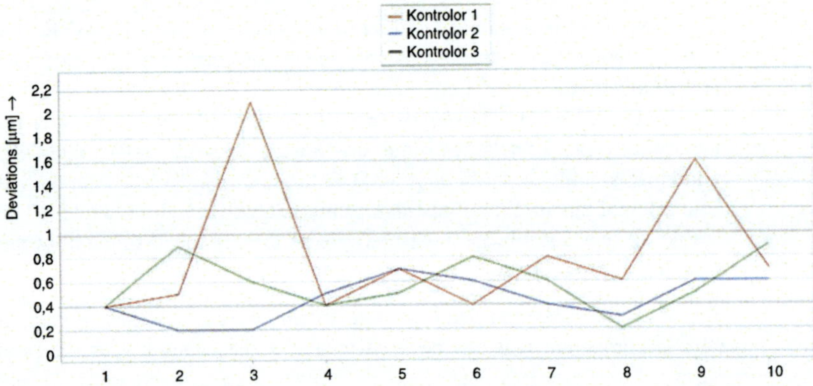

Fig. 7 Deviation diagram

inspectors and provides evidence of any outliers. The analysis of this graph shows that there were small differences between the inspectors.

Figure 7 shows a diagram of deviations between inspectors' individual measurements. The deviation diagram showed a visible difference among the inspectors. The greatest deviations in the measurements were recorded by inspector 1.

Figures 8, 9, and 10 show boxplot graphs for individual inspectors. In the graphs, we can see whether the measurements and variability between the inspectors are accurate. The red dots show the mean data values of each inspector. The error lines show the scatter of values in individual measurements. This was confirmed by the finding that the highest quality measurements were achieved by inspector 1.

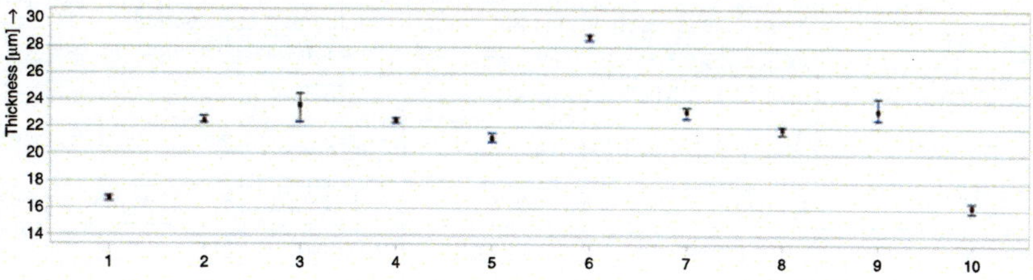

Fig. 8 MSA boxplot for inspector 1

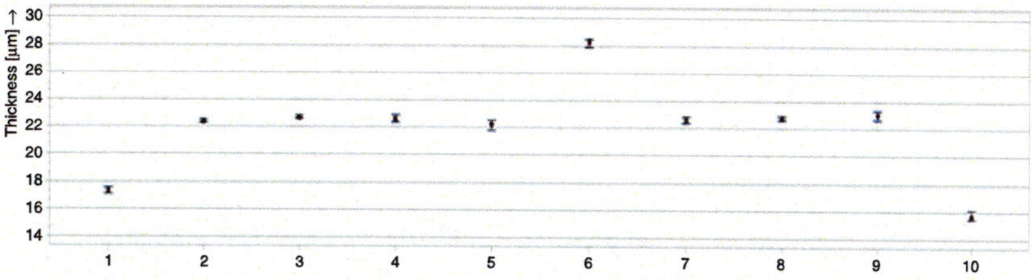

Fig. 9 MSA boxplot for inspector 2

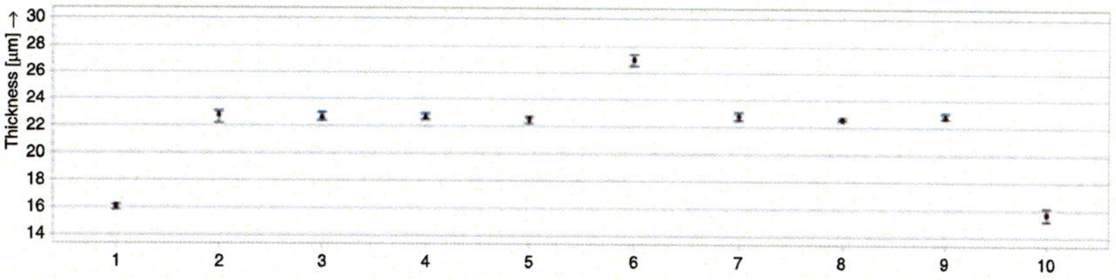

Fig. 10 MSA boxplot for inspector 3

4 SPC Analysis (Table 2)

Based on the measured data, we can say that the level of performance of the process in January was low. This is characterized by the values of the process performance indices Pp and Ppk (0.75 and 0.45, respectively). Based on the evaluated process capability indices Cp and Cpk, we can say that the process is stable but outside the control limits. Similarly to January, the measured data show that the level of performance of the process in February was low. This is characterized by the values of the process performance indices Pp and Ppk (1.06 and 0.64, respectively). Based on the evaluated process capability indices Cp and Cpk, we can say that the process is stable (Cp = 1.71) but is poorly centered (Cpk = 1.03). Based on the data measured in March, we can say that the level of performance of the process in March was low. This is characterized by the values of the process performance indices Pp and Ppk (0.67 and 0.66, respectively). Based on the evaluated process capability

Table 2 The results of the SPC analysis for the measured period

Final values			
	Measurements in January	**Measurements in February**	**Measurements in March**
X'	19,49 μm	19,53 μm	22,42μm
Median	20,3μm	18,8μm	21,8μm
R	14,2μm	12,6μm	39,3μm
S'	0μm	0μm	0μm
σ	3,32μm	2,35μm	3,72μm
Min	12,6μm	14,7μm	16,7μm
Max	26,8μm	27,3μm	56,0μm
Distribution	Normal	Normal	Normal
Calculation method	M1,5 [σ]	M1,5 [σ]	M1,5 [σ]
Calculation method Cp	M1,4 [R'/d2]	M1,4 [R'/d2]	M1,4 [R'/d2]
Pp	0,68≤**0,75**≤0,83	0,96≤**1,06**≤1,17	0,61≤**0,67**≤0,74
Ppk	0,41≤**0,45**≤0,50	0,58≤**0,64**≤0,70	0,60≤**0,66**≤0,73
Cp	1,44≤**1,59**≤1,75	1,54≤**1,71**≤1,87	1,07≤**1,18**≤1,30
Cpk	0,86≤**0,96**≤1,05	0,93≤**1,03**≤1,13	1,06≤**1,17**≤1,29

indices Cp and Cpk (1.18 and 1.17), we can say that the process is stable but within the control limits.

Control Diagrams (Figs. 11 and 12)
In the control diagrams for the month of January, we can see plotted representation of the statistical regulation of the production process. As can be seen in the control diagram of means, several values are measured outside the control limits. These findings are confirmed by the values of the performance and process capability indices.

As can be seen in the control diagrams in Figs. 13 and 14, the production process is in better condition than it was in January. The measured values are within the control limits. Several trends can be observed in the control diagrams. The level of process centering is low.

As can be seen in the control diagrams in Figs. 15 and 16, almost all values are measured within the control limits. Two values are slightly outside the control limit and one value is attributed to measurement inaccuracy.

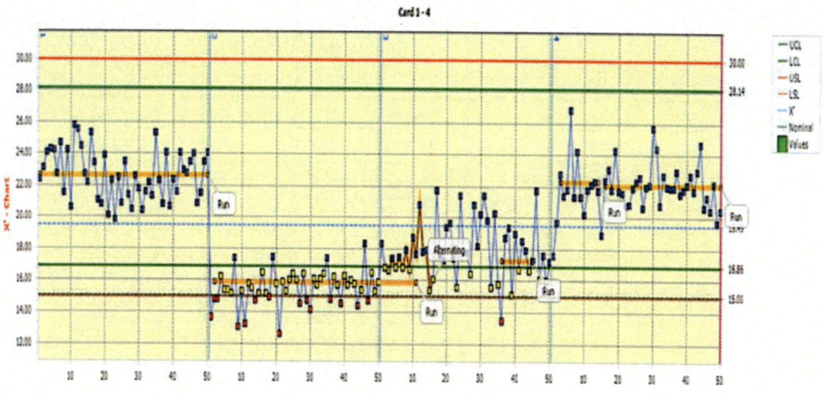

Fig. 11 X-graph depicting measurement in January 2022

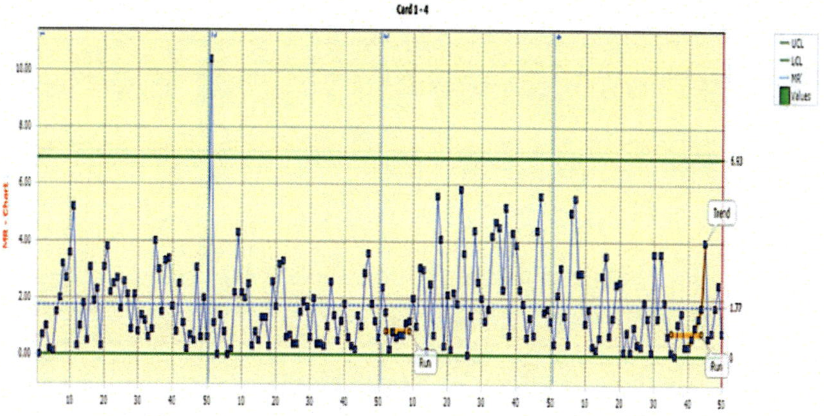

Fig. 12 MR-graph for January 2022

5 Conclusion

Intensive growth in the automotive sector places great demands on manufacturers. Mainly on the production of automotive components, for which we use experimental methods in production technologies, such as measurement system analysis and statistical process control. Analysis of the measuring system is a potential technique to perform an inspection of the observed measuring device with good results for all parameters that take into account the center position of the measurement, coupled with the statistical process control as a feasible method of developing quality in vehicle production. Using control diagrams, the quality assurance department will continuously monitor the process of carrying out design activities, reducing duplication and rework.

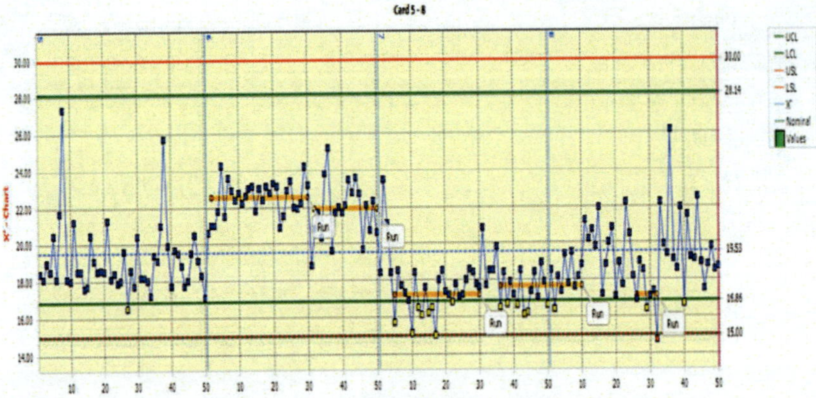

Fig. 13 X-graph plotting measurements done in February 2022

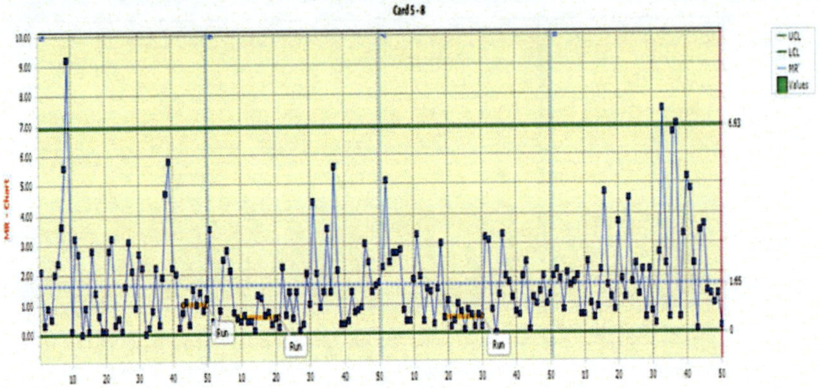

Fig. 14 MR-graph for February 2022

The aim of the work was to determine the analysis of the measurement system and the statistical system of production process management using the PALSTAT CAQ software. A part of the BMW car lock was used as the analyzed component. The analysis of the measurement system found that the given measurement system is satisfactory and applicable to the subsequent statistical management of the production process capability. During the statistical control of the production process, performance and capability indices of the production process and control diagrams were evaluated. Based on the evaluated indices and control diagrams, we can say that the production process is not in good shape and should be improved.

The following research is presented under the title experimental analysis of the influence of technological factors on the properties of the layer created by cataphoretic painting. Under this term, we can imagine the monitoring of selected factors of cataphoretic painting during the entire process. With the help of testing

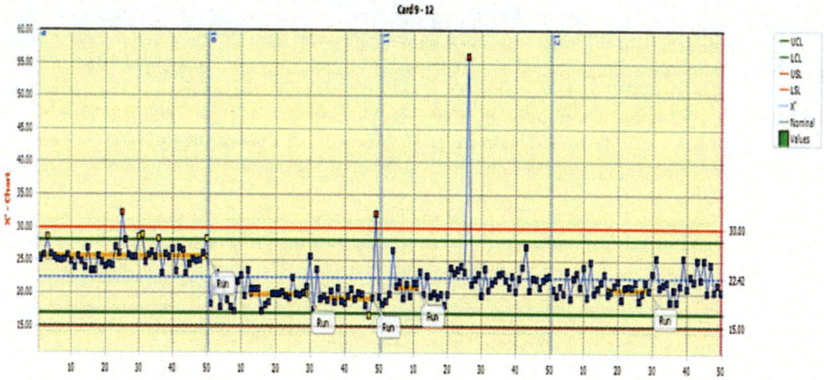

Fig. 15 X-graph plotting measurement done in March 2022

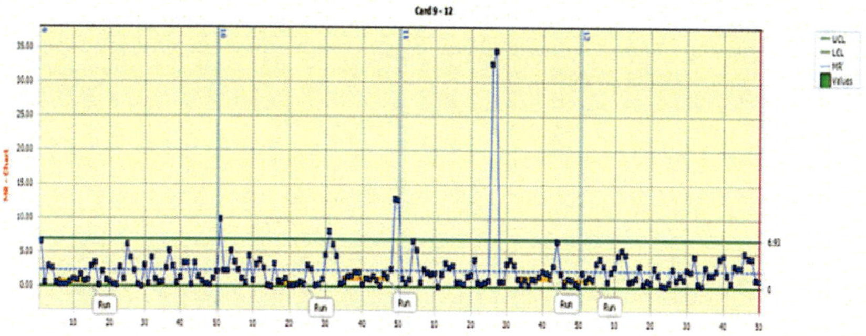

Fig. 16 MR-graph of measurement from March 2022

equipment, we will test the thickness of the created layer, adhesion to the surface, impact test, and conical bending test on the samples. By evaluating the measurement results, we will find out the influence of factors on the properties of the cataphoretic coating. If the results are interesting for the company, we will continue with process optimization.

Acknowledgments This paper has been elaborated in the framework of the project KEGA no. 063TUKE-4/2021.

References

1. Cem, U. C. A. N., Bora, M., & Coban, O. (2019). The effect of Cataphoresis coating on fatigue behaviour of control arms under corrosion environment. *The Eurasia Proceedings of Science Technology Engineering and Mathematics, 7*, 212–219.
2. Zubrzycki, B. (2017). Zabezpieczenie na długie lata. *Lakiernictwo, 1*(105).

3. Romano, A. P., Fedel, M., Deflorian, F., & Olivier, M. G. (2011). Silane sol–gel film as pretreatment for improvement of barrier properties and filiform corrosion resistance of 6016 aluminium alloy covered by cataphoretic coating. *Progress in Organic Coatings, 72*(4), 695.
4. Skotnicki, W., & Jędrzejczyk, D. (2021). The comparative analysis of the coatings deposited on the automotive parts by the Cataphoresis method. *Materials, 14*(20), 6155.
5. Bajat, J. B., Miskovic-Stankovic, V. B., Vesna, B., & Drazic, D. M. (2007). Adhesion of epoxy cataphoretic coatings on Zn alloys. *Journal of the Serbian Chemical Society, 72*(12), 1383–1392.
6. Szmigielski, M., et al. (2017). Effect of temperature cataphoretic process on selected properties paint coatings and their corrosion resistance. *Przemysl Chemiczny, 96*(4), 863–865.
7. García, S. J., & Suay, J. (2009). Optimization of deposition voltage of cataphoretic automotive primers assessed by EIS and AC/DC/AC. *Progress in Organic Coatings, 66*(3), 306–313.
8. Rossi, S., & Calovi, M. (2018). Addition of graphene oxide plates in cataphoretic deposited organic coatings. *Progress in Organic Coatings, 125*, 40–47.
9. What is cataphoresis : https://www.faiauto.com/what-is-cataphoresis/ 2022/4/15.
10. SMK Automotive Industry; Methods of painting, Stages of Car Production, Characteristics of Water solutions for Cataphoresis, Available From: [https://automotiveindustry.ru/en/smk/] 2022/04/16.
11. Matuška, M., & Kučerová, M. (2018). Moderní trendy v kataforéze. In *Povrchové úpravy [online]*. Praha, Available From: http://www.povrchoveupravy.cz/PDF/2-2018/PU_2018-02_15-18.pdf 2022/4/15

An Effective Tool for Increasing the Company's Production Process – Application of Innovative Planning Methods

Annamária Behúnová and Marcel Behún

1 Introduction

By planning, we mean following the path to the final goal, displaying the sequence of events predicted reasonably. We were planning in the present to decide what to do in the future. It includes anticipating the consequences and events that can affect the business. It requires creativity and analysis in defining business opportunities and constraints. It provides an internal decision-making structure for the company as a whole. It requires thinking about shaping the future of the company instead of expectations [1, 2].

Important indicators include the change associated with the company's activity, development, structure and resources. With the help of planning, managers determine the parameters of the company's desired performance and individual coordinate components and individual employees to meet the given goals [3].

Planning is focused on determining the organization's future state and ways to achieve them. It determines the goal concerning resources (financial, personnel, technical, etc.) and sets the paths and specific activities to achieve these goals. In the case of business entities, it is necessary to take into account the behaviour of competition. If the organization is in an unsatisfactory situation, it may result in incorrect planning and incorrect determination of objectives or procedures [4, 5].

The planning process consists of three basic phases [3]:

1. Analytical phase – consists of internal and external analysis. By internal analysis, we mean all functional areas of the company (production, production resources,

A. Behúnová (✉) · M. Behún
Technical University of Košice, Faculty of Mining, Ecology, Process Control and Geotechnologies, Institute of Earth Resources, Košice, Slovakia
e-mail: annamaria.behunova@tuke.sk; marcel.behun@tuke.sk

L. Knapčíková, D. Peraković (eds.), *7th EAI International Conference on Management of Manufacturing Systems*, EAI/Springer Innovations in Communication and Computing, https://doi.org/10.1007/978-3-031-22719-6_11

marketing, supply, personnel and financial security, organizational structure, etc.). The main goal should be the company's performance compared to the competition, the efficiency and effectiveness of production resources, the company's priorities and the development of business results. This can be identified by identifying the strengths and weaknesses of the business. External analysis is used to determine the external effects on the company. These are the effects of political events, the state of the economy, the labour market and demographic developments. The analysis task is to identify opportunities and threats to the company.
2. Conceptual phase – used to develop variants of the plan. Variant planning is the company's response to the possible risks of implementing the plan caused by changes in production and input prices, variability in demand and inflation. It is recommended to create three variants: optimistic, realistic and pessimistic.
3. Implementation, control phase – consists of transforming intentions into reality, while it is necessary to evaluate or re-evaluate the plan's implementation preliminarily. There are often deviations from the plan that need to be minimized. The corrective measures taken should ensure that the organizational objectives are achieved.

There are many technical and system changes in the manufacturing industry. This also changes the requirements of the market for the companies. Production planning and control respond effectively to these internal and external changes, ensuring faster responses and better control over performance and delivery [6].

When planning, managers must make decisions that relate to its core elements – goals, procedures, resources, tasks and control. If they are to be effective, they are related to the basic concept that governs human activities (Fig. 1). Based on this, all managers should be involved in planning at all levels of management. Each manager must set a goal, make decisions and achieve the desired goal. Planning has a direct impact on [7, 8]:

1. Efficiency of organizational activities
2. Risk reduction
3. Development of managers and organizations
4. Level of integration and coordination of efforts
5. Organizational performance

Goal setting – belongs to the initial and critical elements of planning, which defines and specifies the achieved future state, and determines priorities [9].

Creating a plan – pays attention to achieving the best and most effective goal. It depends on the type of plan, the information obtained, the quality and who will participate in the planning cycle [9].

Monitoring of plans – solves the current state of whether the planned goals are being met and whether something needs to be adjusted [9].

Result or control – belongs to the final phase, evaluates the result and to what extent the goals have been met [9].

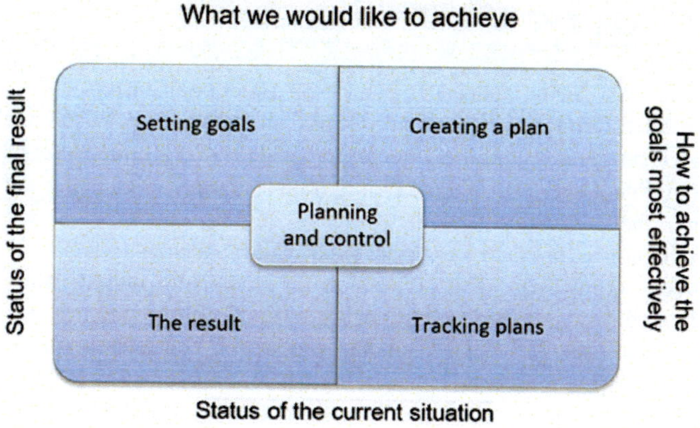

Fig. 1 Phases of the planning cycle [9]

To meet market requirements, companies prepare a Master Production Schedule (MPS), which specifies the number of individual products scheduled in a certain time horizon and the sales program to sell the products. The main goal of production planning and management is to ensure the production of products and semi-finished products according to the master production plan, which is in line with the implementation of other measures [6].

The production planning and control: PPC system uses a hierarchical planning process that helps the manager understand and manage operations. A common starting point for production planning is the main production schedule, from which the requirements for material, parts, machines and work are derived. If the requirements are disproportionate, they can be adjusted. A high-level plan determines the context in which lower-level plans appear [6].

The goal of PPC is to plan and manage production so that the company meets production requirements as efficiently as possible, producing what the market requires in the required quality, volume, and time and at a minimal cost. The PPC system contains all the tools and processes to achieve this goal. One of the key elements in operations management research is the adaptation of PPC to the production system, which determines the production company's efficiency, profitability and long-term capability [10].

Production planning and control technology must be changed to a properly designed process in which the consequences of changing system parameters are understood. Better conceptual understanding and improved PPC operational performance are important factors in managing the rest of the organization. Companies must plan and coordinate all functional stages associated with the introduction of the product, from the initial design of the product, through production, distribution, warranty and post-warranty service to eventual recycling. It is important to respond better to external changes, including competitive measures and changing market demand. For companies, this means meeting newly perceived needs and adapting systems without high costs [10].

1.1 Efficiency of the Production Process

Production can be defined as converting production factors into economic goods that pass through consumption. Goods are referred to as physical commodities (produced for consumption) that meet the needs of consumers. Services are required intangible assets [11]. Production is transforming and adapting the resources that enter the production system to create tangible goods or services. The aim of production is products or services that can be realized on the market and obtain revenue from them. The conversion of inputs into outputs must take place as efficiently as possible by introducing optimal consumption of production inputs, reasonable costs, the most suitable production procedures and compliance with safety and environmental conditions [12].

From a technical point of view, production is described as an operation with a purposeful technical interconnection of all production factors. It uses production equipment with or without a worker's participation and converts the material into tangible goods. There is no material transformation when providing services [12].

From the economic and social point of view, the production goal is to achieve a state in which all production resources are used efficiently. Efficiency in a broader sense means using resources in production in a way closest to the goal of business and profit generation. To a large extent, competition affects producers to use production factors as efficiently as possible and produce a certain amount of products with the lowest possible consumption [11]. In practice, the productivity indicator, also determined as the profitability indicator, is used to evaluate the use of production factors, which means the ratio of output to consumed factors of production. Sometimes it is necessary to assess the outputs from a non-financial point of view. There may be a case where the output is financially lower than the price of the factors of production consumed [11, 13].

The benefits of an efficient production plan [14]:

1. Reduced labour costs by eliminating lost time and improving flows
2. Reduced inventory costs by reducing the need for safe or excess inventory during production
3. Optimal use of the equipment and increased capacity
4. Improved delivery of products and services

The production plan must be solved well in advance of production to ensure a continuous flow of work during its development [14].

2 Methodology

Several methods and techniques can be used in planning: analysis, synthesis, induction, deduction, simulation methods, economic-mathematical modelling and others. Planning techniques make it possible to visualize planning tasks and monitor them. Methods and techniques are used in the planning process [3]:

- In the analysis and preparation stage of the plan
- In the process of creation
- When monitoring the progress of implementation
- In the control stage

2.1 *Enterprise Resource Planning*

Enterprise Resource Planning (ERP) is a system or software used to plan and manage an organization's processes, from manufacturing/services to financial or business. It is used to automate and simplify individual activities in the company and various industrial areas and supports the connection of modules [15]:

- Financial and managerial accounting
- Production
- Human resources
- Order processing
- Data services and others

There are three main types of ERP systems: On-premise software, Cloud software or "Hybrid" software.

On-premise (local) software is implemented in office space and used on the company's computers and servers.

An organization uses cloud software to store and work on any device with an Internet connection. The software provider allows for updates, various pieces of training and ongoing support.

"Hybrid" software combines cloud and local software with a difference in service delivery [15].

An ERP system is considered a type of enterprise application that improves performance and meets the needs of an organization. It varies depending on customers' size, functions and needs, which helps eliminate costly and time-consuming processes [15].

2.2 *Material Requirements Planning*

Effective management of raw materials and components flow from the supplier through production to the customer is challenging. In the event of poor management, there will be a possibility that the part or raw material will not be available. This will disrupt the production process and fulfil the planned orders and deadlines. As master production plan records are combined with inventory data, components and parts can be issued and replenished. This process is called Material Requirements Planning (MRP) [16, 17].

The MRP method is used to calculate the components and materials that will be needed for production again. It is an inventory planning and management system that takes into account information about current and future orders [16].

2.3 Just in Time

One of Just in Time's (JIT) priorities is minimizing inventory and shortening production lead times. The system is mainly applied in mass or large-scale production, but some elements can also be applied in production processes with a lower degree of repeatability. The most significant characteristic is the pressure on very low stocks of work in progress and finished products. The goal of JIT is to involve employees in the process, constantly improve it, solve problems and produce suitable products in appropriate quantities before they reach the customer. Because the system works with low stocks, the material is delivered to the workplace shortly before processing and passes in small batches. The machine operator is responsible for handing the functional product to another workplace. In the event of a defective product, the worker is obliged to stop the machine. The advantage of this system is the simple and fast identification of the worker who processed the material, quickly discovered the cause, took measures and prevented errors to a greater extent [18].

On the other hand, stopping machines is seen as a space for a natural process of improvement. When the machine is not in operation, the worker can devote himself to maintenance, repairs, design and remedial measures, thus creating space for the worker to be trained. In the JIT system, it is necessary to prioritize preventive and predictive maintenance to prevent machine failures, with the possibility of not disturbing the planned synchronization of workplaces. In this case, complete, comprehensive, productive maintenance is applied – Total Productive Maintenance (TPM) as a group technology with a cellular organization, which allows rapid implementation, transfer of responsibility for the workplace, including machine condition, routine maintenance and simple diagnostic procedures [18].

2.4 Production Planning in the Company

The production process in the company begins by entering a production order. The request for production can be submitted by email by an employee from the department of sales, development, realization of objects, warehouse manager and executive director [19].

Table 1 Production planning in the excel program

		Month				
		Week				
		Mon.	Tue.	Wed.	Thu.	Fri.
		Date	Date	Date	Date	Date
Product name:	Operation					
	Operation					
	Operation					
	Operation					

The production order contains:

- The name of the product or semi-finished product
- Number of pieces
- Creation date, end date
- The name of the final recipient, or section
- Notes
- Name of the guarantor of the exit control
- Total time worked

The responsibility for planning lies with the production director or employee, who has an overview of production and knows the duration of individual operations (production of semi-finished and finished products).

The data needed for production planning are the working time of individual production operations, an indication of the approaching completion date, and the connection between orders and the order of production operations.

Production planning is implemented and designed in the company in Excel and Google Docs tables (Table 1). The production plan contains currently offered products with individual operations with a total annual scope. The planning of production operations is clear due to the colour coding of each operation and the expected duration. Each process is mapped in order, is simple and contains all the necessary information.

The employee responsible for planning works with the actual state of the material and semi-finished products in the warehouse. Depending on the stock status, it orders material to carry out the required number of pieces from external suppliers. After the production is planned, the delivery date is confirmed between the customer and the production. Subsequently, the production order with the necessary data and the end date is entered into the company system.

A bill of material (in the company called kusovnik) is a document that contains a list of parts, materials and raw materials needed to produce a given product. The company uses a multi-level bill of material (Table 2). It shows assemblies and components arranged in a top-down manner. The parent components are at the top and have their child components.

The task of the warehouse worker is to order, store, and prepare material and components for production via a bill of material or pack the components

Table 2 Bill of material of the company

Product name	Drew:		
Name: Bill of material	Date:		
Drawing number:	Purpose:		
Name	*Drawing number*	*(number)*	*Note*
Main part			
Material			
Material			
Material			

and technical documentation and send them to an external company for partial production.

The technical documentation contains:

- Product description
- Production, construction drawings and their explanation
- Technological procedures
- Bill of material

The production process starts according to the production plan. After delivery and removal of the material, the head of the department assigns the work to the production operator. He is familiar with the production process and works on the production order. He completes the product according to the documentation and sends it for inter-operational control. During series production, a record card is kept for the products, which contains the number of the production order, the stage number, the number of the printed circuit board, the date of the beginning and end of production, the type of operation and the name of the operator.

Interoperation Control

Interoperation control is used to control compliance with production procedures – technical documentation. Depending on the product type, the authorized electronics perform several inspections in the production process. Interoperation control is divided into visual control and functional control (recovery/adjustment/calibration). Any employee of the production department can perform a visual inspection. This inspection is focused on the completeness and integrity of semi-finished products. Operations related to operational control, recovery or adjustment may be performed only by an employee with the inclusion of electronics – mechanics, by the valid Decree no. 508/2009 Coll. (minimum §21) on the relevant test device. Based on the revitalization regulations, he will perform the revitalization/adjustment of the product, paint all printed circuit boards with a protective varnish and mark the product with his brand. In this way, the product is ready for completion. If a non-compliant product is found during the in-process inspection, it is returned to the operator for repair and elimination of defects.

Completion of Products
Based on the production process and technical documentation, the production operator completes the product in a condition to meet the final inspection criteria. It is mainly the assembly of semi-finished products into products, assembly into boxes, etc.

Checkout
The output control technician verifies the functionality and parameters of the product according to the relevant related control procedure – visually and by measurement. It checks the materials' correctness and makes a record of the test in the report from the final inspection and archives. After the final inspection, the products are marked with a serial number or QR code.

3 Results

Based on the analysis of the company's production process and planning process, but also the basis of experience and knowledge of employees in individual operations, it is possible to work on streamlining production processes, specifically updating bill of material.

A proposal for streamlining the flow of material and components is applying the material requirements planning method – MRP. This method is used to calculate the components and materials that will be needed for production again. This inventory planning and management system take into account information about current and future orders.

MRP system:

1. Help to plan purchasing, delivery and production activities without delay
2. Will have raw materials for production and products for the customer
3. Meet the appropriate storage condition of the products

The goal of material requirements planning will be to manage orders so that delivery takes place just in time (Fig. 2). It will be possible to order parts in advance to keep the appropriate quantity in stock, shortening the preliminary production time (just in time method), which will be available for repetitive production. The MRP system can work with the JIT method at the same time.

Once the technologist has been hired, another possible proposal would be to implement an enterprise resource planning (ERP) system. It is an information system ABRA Gen, which will help streamline the company's operation and has applications in all business areas (Fig. 3).

ABRA Gen covers several business areas:

- CRM/customers
- Purchase
- Warehouse
- Production

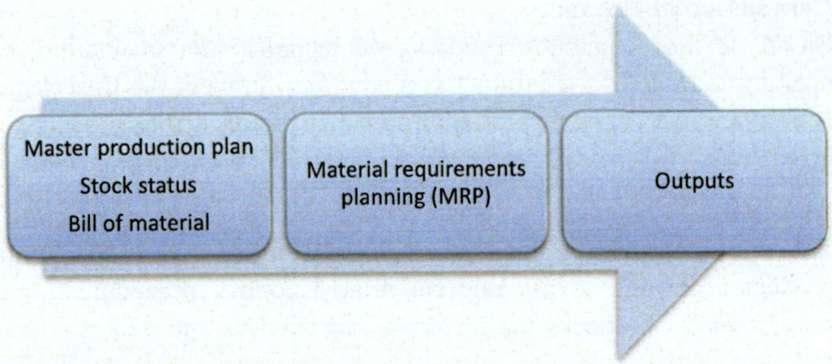

Fig. 2 Application of the MRP system [16]

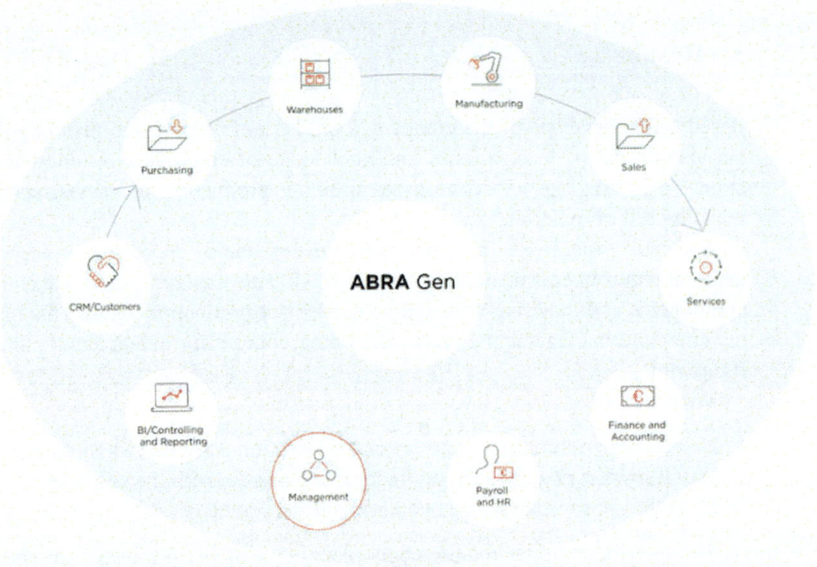

Fig. 3 A schematic of areas covered by ABRA Gen ERP [20]

- Sales
- Provision of services
- Finance and accounting
- Wages and human resources
- Management and administration
- BI/controlling and reporting

The ABRA Gen all-in-one is a powerful information system that ensures a fast flow of documents and information, helping automate and analyse all the necessary data in the company. The system's advantage is adapting to the needs of

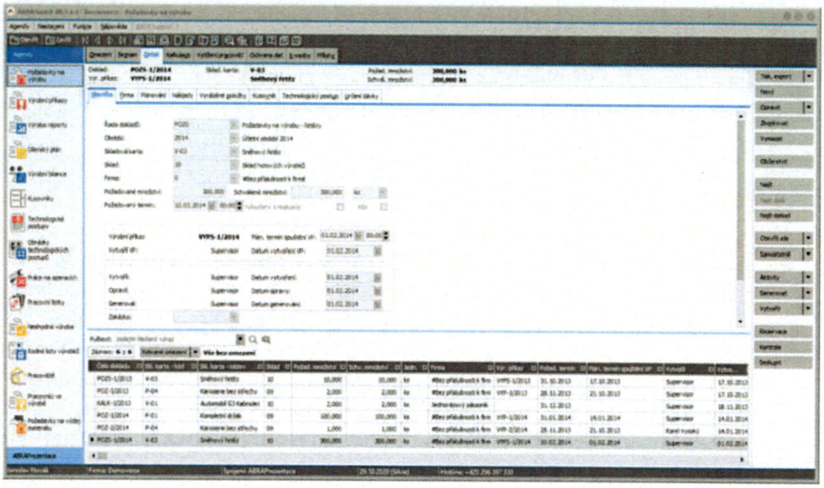

Fig. 4 Information system ABRA Gen [20]

a specific company with the most modern way of connecting using the application programming interface (API). Thanks to the API, data from all applications stored in one place meet in the information system, e.g. the data from the refuelling card on the amount of fuel on the business trip will be entered into the information system [20].

3.1 Use of the ABRA Gen Information System in the Production Process

The information system can monitor and control the production process through production orders, also used by the researched company (Fig. 4). It contains all the information about the product and its production history linked to the given production order [20].

The system uses methods and work queues to help manage workplaces and individual operations to streamline production process planning. After their implementation, it evaluates graphs or statistical data [20]. With the help of defined standards in repeated production, it can streamline the assembly of products composed of the same or similar components, calculate the product price and compare it with the real price of production [20]. It uses supply chain management (SCM), which provides information on the state of the material in the warehouse and draws attention to the minimum and maximum stocks [20].

Implementing the system would greatly benefit the company in terms of managing all processes "in one" to connect all work sections. This would make the operation of the whole company more efficient and better.

4 Conclusion

The main goal of the presented work was to streamline the planning of production and production processes in the company. Based on knowledge and information from the literature, based on an analysis of the company's current state and using methods of material management and enterprise resource planning, the company created a proposal to streamline the production process, namely updating bill of material through the application of information system. With the possibility of further development, the all-in-one information system ABRA Gen was designed to standardize production and support processes and unify all work units of the organization. The system will help provide a standardized time reporting and expense tracking platform and significantly increase an organization's ability to store the necessary data. The advantage of implementing the ABRA Gen information system is the support of the fulfilment of business goals, the adjustment of the functioning of processes, the clarity and speed of obtaining information, the optimization of stocks and others. Thanks to this, the company will be able to minimize errors and reduce downtime in one system. Gradually, the need to transfer information and fill in paper documentation will be lost, thus reducing production preparation time. Based on the streamlining of several processes within the company's operation, other areas of research are opening up, for example, research on the impact of information technology on reducing the company's variable costs or increasing the productivity of employees.

References

1. Vaishnavi, N. Planning in management. *Business management ideas*. https://www.businessmanagementideas.com/management/planning-management/planning-in-management-2/12196
2. Murgaš, J. (2000). Plánovanie marketingu a výroby. Agroinštitút, .
3. Grznár, M., et al. (2009). Firemné plánovanie. Sprint dva, Bratislava.
4. Murgaš, J. (2001). Plánovanie: teória a metodológia. Agroinštitút, .
5. Veber, J., et al. (2007). *Management základy, prosperita, globalizace*. Management Press.
6. Bonney, M. Reflections on Production planning and Control (PPC), https://www.scielo.br/scielo.php?pid=S0104-530X2000000300002&script=sci_arttext
7. Bělohlávek, F., et al. (2006). *Management*. Computer Press.
8. Harausová, H. (2005). Manažérske funkcie. Metodicko-pedagogické centrum, .
9. Žurková, H. (2007). Plánovaní a kontrola klíč k úspěchu. Grada, .
10. Oluyisola, O. E., et al. (2020). *Smart production planning and control: Concept, use-cases and sustainability implications*., https://www.mdpi.com/2071-1050/12/9/3791.

11. Keřkovský M., et al. (2012). Moderní přístupy k řízení výroby. Nakladatelství C. H. Beck, Praha.
12. Makovec J., et al. (1993). Organizace a plánovaní výroby. Vysoká škola ekonomická v Praze, Praha.
13. Knapčíková, L., et al. (2021). Challenges of industrial engineering, management and ICT. *Wireless Networks, 27*(3), 1557.
14. bdc: What makes a good production plan? Discover the five key factors of an effective production plan. https://www.bdc.ca/en/articles-tools/operations/operational-efficiency/production-plan-top-tips-improving-operations?fbclid=IwAR15UEGp-K5LzFCS4dpb9URrM51_M4QkBq09eKiknQG37jyNa707UKHsioo
15. QAD: Definition of Enterprise Resource Planning (ERP), https://www.qad.com/what-is-erp
16. Thorat, S. Material requirements planning – input, output, objectives, advantages, https://learnmech.com/introduction-to-material-requirements/
17. Straka, M., et al. (2018). Utilization of computer simulation for waste separation design as a logistics system. *International Journal of Simulation Modelling, Rakúsko, 17*, 583.
18. Nenadál, J., et al. (2008). *Moderní managemet jakostí*. Managemet Press.
19. Mandičák T., et al. (2021). Impact of information and communication technology on sustainable supply chain and cost reducing of waste management in Slovak Construction. *Sustainability, Bazilej Švajčiarsko*.
20. ABRA. (2022). *ERP system ABRA Gen*, https://www.abra.eu

Quality Management of the Process of the Complex Parts Control

Yuliia Denysenko, Justyna Trojanowska, Vitalii Tsarytsyn, and Jozef Husár

1 Introduction

The rapid progress of industry taking into account waste-free "life cycle" requires strong and lightweight constructions without defects [1, 2]. The design of such structures becomes possible using modern software [3, 4]. Optimization of the shapes of parts that accept heavy loads leads to the emergence of complex parts. The manufacture of such parts is possible when using multi-axis machining, as well as additive technologies. However, there are questions about the surface accuracy control of such parts [5], which directly depends on the control methods and selection of control and measuring devices, etc.

A significant number of such parts are included in the designs of aircraft [6] and cars [7, 8]. Today, these industries are constantly evolving [9]. The study of the main tasks of controlling such parts requires an analysis of the requirements for controlled parameters.

It is essential to understand that for these parts, the main thing is to ensure the dimensional accuracy of the surfaces, the accuracy of the shape of the treated surfaces, as well as the accuracy of the location of the surfaces relative to each other. Because it significantly affects the assembly process of the product and its operation.

Y. Denysenko · V. Tsarytsyn
Sumy State University, Faculty of Technical Systems and Energy Efficient Technologies, Sumy, Ukraine

J. Trojanowska (✉)
Poznan University of Technology, Faculty of Mechanical Engineering, Poznan, Poland
e-mail: justyna.trojanowska@put.poznan.pl

J. Husár
Technical University of Kosice, Faculty of Manufacturing Technologies with a seat in Presov, Presov, Slovak Republic

L. Knapčíková, D. Peraković (eds.), *7th EAI International Conference on Management of Manufacturing Systems*, EAI/Springer Innovations in Communication and Computing, https://doi.org/10.1007/978-3-031-22719-6_12

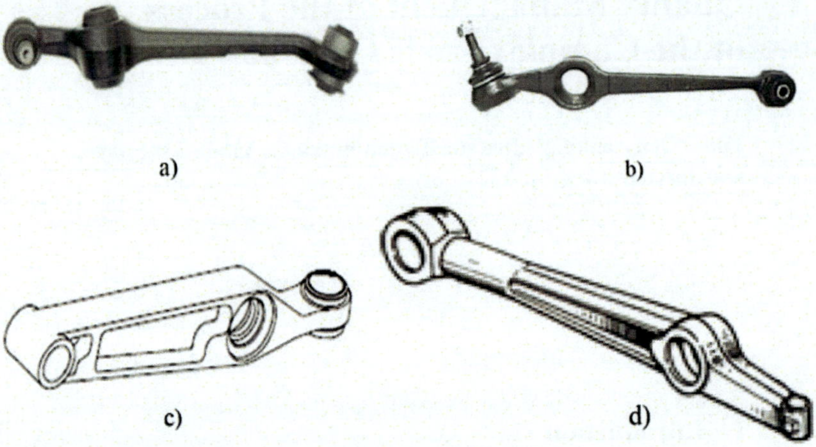

Fig. 1 Examples of complex parts – swing arms of: (**a**) Hyundai Accent, (**b**) LIFAN LF-7162 Solano, (**c**) CHANA Benni, (**d**) Mercedes E200 [10]

Therefore, analyzing the nomenclature of the elementary surfaces of parts of which they consist, as well as the development of appropriate classifications, is a topical issue.

The complex parts include the following: parts of aircraft, light aircraft housings, and molds for workpieces. The automotive industry includes the following types of parts: levers, connecting rods, forks, brackets, and others. The complex parts are shown in example of lever-type parts in Fig. 1.

To obtain information on the volume of complex parts author [10] analyzed the specifications of the mechanisms of cars of different years of manufacture. According to [10], in some mechanisms of modern cars, the number of parts of complex shape reaches 26% and the number of complex parts gradually increases [10, 11]. Most of them are contained in the following assembly units: steering mechanism (26%); engine suspension (25%); rear and front suspension (21% and 16% respectively); engine (20%); clutch mechanism and gearbox (about 15% each).

Complex parts are characterized by the complexity of the setup charts and insufficient tool availability. This is explained by the complex spatial arrangement of the elementary surfaces of which they consist. When controlling the parameters of such parts, there are difficulties in choosing control and measuring tools and providing the necessary measurement techniques. In this case, in addition to providing measurements with a given accuracy of products, it is important to ensure:

- Optimal application for the conditions of a given production system of the most acceptable means of control and measurement, which may consist of universal, standardized, progressive and automated, as well as a set of measuring tools.
- Improvement of methods of control and measuring processes of production.

- Reducing the complexity of control and measurement techniques.

Implementation of ISO 9000 series standards in enterprises requires monitoring and evaluation of process quality [12, 13]. Controlling and monitoring represents an important role in the quality management system [14]. Lasinska [15] admits that at certain stages of a project, traditional project management methodologies become simply insufficient. In those times the above problems significantly limit the quality of the process of controlling the accuracy of complex details. Therefore, today's urgent task is to improve the quality of the process of controlling the accuracy of complex parts sizes (including the accuracy of sizes location, etc.).

The work aims to increase the efficiency of the production of complex parts by making decisions based on assessing the quality of their control processes.

One of the ways to solve this problem is to develop a method for assessing the quality of the control process of complex parts.

2 Literature Review

Although the object of the control process is complex parts, they consist of a large number of elementary surfaces (planes, ledges, cylindrical or conical holes, etc.), which can be mutually located in different planes and at an angle [16]. Ivanov et al. proposed 12 classification attributes of fork type parts, main of it: (1) by purpose, (2) by form of part, (4) by availability of functional elements, (6) by arrangement of the datum surfaces, etc. [11].

According to OKRB 500-94 [17], the parts shown in Fig. 1 are included in class 74. In addition, in the automotive industry there are classes: class 71 – parts – rotating bodies such as rings, discs, bushings, shafts, axles, rods, etc.; class 72 – details – bodies of rotation with elements of gearing; class 73 – parts – non-rotating body, support, capacitive, etc.

Ning et al. [18] proposed the following syntax pattern recognition of complex parts: (1) joint-edge; (2) parallel; (3) vertical; (4) coplanar; (5) intersect; (6) projection; (7) symmetrical; (8) deviate; (9) distance; and (10) coaxial.

Analyzing the proposed classifications of complex parts, we can conclude that manufacturing and controlling such surfaces consist of ensuring and measuring the accuracy of dimensions and location of elementary surfaces, either in complex or separately.

The choice of measuring tool has an impact on the quality of the control process of complex parts. Known classifications of measuring instruments subdivide them by [19]:

- Degree of universality (universal, limited universality, special)
- Degree of automation (mechanized, automated)
- The nature of contact with measuring surfaces (contact, non-contact)
- Method of measurement (absolute, relative)
- Purpose (working, reference, industrial (technical), laboratory)

Table 1 Typical nomenclature of measuring instruments

Universal	Special	Automated
Lever bracket Micrometer Depth gauge Indicator Calipers Template	Special measuring and measuring device Caliber-stopper Caliber bracket	Conventional coordinate measurement machine Optical measuring systems

The study of modern experience allows us to present in Table 1 a standard aggregate nomenclature of measuring instruments according to the classification.

Also, the analysis of world experience [20–23] allows to identify the obvious shortcomings of the choice of measuring instruments in the development of the process of control of complex parts in the production:

- The purpose of measuring tools is carried out separately for each operation, which is inefficient for multiproduct manufacturing.
- The appointment of universal and standardized measuring tools is ineffective in the presence of advanced and automated control tools.
- High complexity of control in the absence of automated tools for the formation of control results.
- Lack of professional development of the controller due to simple control and measuring tools.

These issues significantly limit the quality of the process of controlling complex parts. And therefore, at an estimation of the quality of the process of complex parts control, it is possible to divide the basic parameters:

1. Fixing the measuring instrument to the parts
2. Level of automation
3. The complexity of control
4. Qualification of the controller
5. The complexity of preparation for control

3 Research Methodology

To assess the level of quality of the process of control of parameters of complex parts, a complex indicator is proposed, which is calculated according to the formula developed using the laws of qualimetry [24]:

Table 2 Score characteristics of evaluation of individual indicators of Qi according to criteria

Number of points for a single indicator of the quality of the control process	Description of compliance with the criteria
10–9	Very good
8–7	Good
6–4	Average
3–2	Satisfactorily
1–0	Unsatisfactorily

$$R = \sqrt[m]{\prod_{i=1}^{m} \frac{\sum_{j=1}^{n} K_j \cdot \gamma_j}{10}} \cdot 100,\ \% \tag{1}$$

where $i = 1 \ldots m$ – number of measuring instruments; $j = 1 \ldots n$ – the number of indicators of the quality of the control process; K – the value of a single indicator of the quality of the control process in the classification group (in points); γ – the weight of the unit quality indicator of the control process.

For clarity of presentation of the complex indicator R it is normalized to the reference value, which corresponds to 100%. In addition, the conversion factor of dimension 100/10 is taken into account, as the maximum value of the unit quality indicator of the control process K_j is 10 points.

The score characteristics of individual indicators evaluation for the quality of the control process, as an example, are given in Table 2. In the future, for scoring individual indicators of the quality of the control process, it is proposed to develop scales for their evaluation using the fuzzy logic methods described in [25].

The value of the single indicator weight of the quality of the control process must also be substantiated. In this case, the expert methods are proposed, namely the expert method of pairwise comparison. This method is a set of mathematical, statistical, and logical methods based on the analysis of information by experts and decision-making.

The algorithm for determining the weight of a single quality indicator is divided into two main stages: the first – the choice of the optimal set of names of features X, and the second – the assignment of a set of numerical values x of each feature.

Determination of the resulting estimates for each of the studied indicators is based on the examination of the algorithm, which is the stages shown in Fig. 2.

At the stage of formation of the set of quality indicators, an expert survey form is formed, which includes a list of quality indicators and expert evaluation (an example is given in Table 3).

The next step is to determine the expert group. It is recommended to involve heads of control departments, controllers, metrologists, as well as researchers or supervisors involved in the study of measurement and control processes in the expert group.

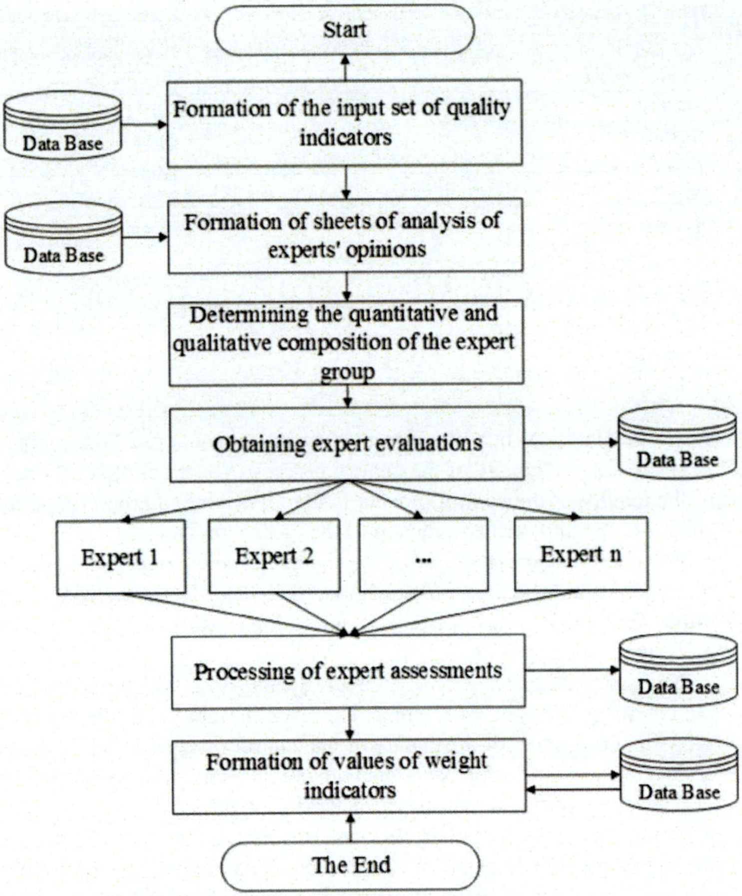

Fig. 2 Algorithm for conducting an expert method of pairwise comparison in determining the weight of quality indicators

Table 3 Expert survey form

		Indicators *i*				
Indicators *j*		1	2	3	4	5
1	Fixing the measuring instrument to the parts	1				
2	Level of automation		1			
3	The complexity of control			1		
4	Qualification of the controller				1	
5	The complexity of preparation for control					1

The number of expert groups is calculated according to the formula proposed by Azgaldov [26]:

$$n = \frac{0.04d^2}{\overline{\Delta q}^2 (1 - \overline{\gamma})}, \tag{2}$$

where d – the scope of the measurement scale; $\overline{\gamma}$ – the value of the fiduciary probability with which the value of the collective expert evaluation is determined will be admissible. Usually the level of significance is chosen and then the confidence probability $\overline{\gamma} = (1 - \alpha) = 0.95$, $\overline{\Delta q}$ assuming the value of the absolute error (Δq) of the collective expert evaluation. It is recommended to take for further calculations $\overline{\Delta q} = 6$ or 7.

After filling in the forms, the experts make the necessary calculations and weigh the opinions of experts according to the method described below.

It is proposed to clarify the weight of indicators in this paper by the method of sequential approximation. In Table 3, the weight values are obtained by double pairwise comparison and sequential approximation. That is, the measurement result in the (w) approximation is defined as the standard weighted average. The initial results are taken as the first approximation. In the second approximation, they are used as the weight $\gamma_j(2)$ of expert opinions. After the next accounting of these weights new results are received and in the third approximation are considered as coefficients of weight $\gamma_j(3)$ of the same opinions of experts again.

The Peron-Frobenius theorem [27, 28] emphasizes that this process is convergent under certain conditions performed in practice. Thus, the weights are close to constant values and this reflects the relationship between the indicators set by the experts of the input data.

The result of the calculation of the j-th indicator for the first approximation $G_j(1)$ (initial result) is determined by the formula [28]:

$$G_j(1) = \sum_{l=1}^{m} K_{jl}, \tag{3}$$

where K_{jl} – the number of benefits of the j-th indicator by one expert ($l = 1 \ldots m$).

The results of the calculation of the j-th indicator in the (w) approximation will be equal to [26, 28]:

$$G_j(w) = \sqrt{[(G_1 (w-1)]^2 \cdot K_{j1} + \cdots + [(G_m (w-1)]^2 \cdot K_{jm}}, \tag{4}$$

where $Gj(w-1)$ – the result of the calculation of the j-th indicator in the ($w-1$) approximation.

The values of the weights in the (w) approximation are determined by the formula:

$$\gamma_j(w) = \frac{G_j(w)}{\sum_{j=1}^{m} G_j(w)}, \tag{5}$$

Refinement of values will continue until the specified accuracy is reached, i.e., until the following condition is met:

$$\gamma_j(w) - \gamma_j(w-1) \le \varepsilon, \tag{6}$$

where ε – the accuracy of the calculations, which is usually approved [27]:

To exclude the case of falling into the group of an incompetent expert, assess the consistency of their opinions on the basis of the calculation of the coefficient of consistency of opinions – the coefficient of concordance Kendel [26]:

$$W = \frac{12\sum_{i=1}^{n}(\gamma_i - \overline{\gamma})^2}{m^2\left(n^3 - n\right)}, \tag{7}$$

where n – number of quality indicators; γ_i – weight of experts for each individual quality indicator; $\overline{\gamma}$ – average weight for all individual quality indicators; m – number of experts.

If the value of the Kendel concordance coefficient $W \ge 0.6$, the agreement of experts is considered acceptable.

4 Results

The result of elaboration of the sheet of analysis of opinions of the member of the expert group is presented in Table 4.

Note. In Table 4 the advantage of the j-th object over the i-th number is denoted by the number 2, the equivalence – the number 1, and the advantage of the j-th object before the i-th number 0.

As a result of calculation of expert estimations the table of weights of indicators of quality of process of control of complex parts is resulted (Table 5).

To draw conclusions about the level of process quality based on a comprehensive quality indicator, it is proposed to use the well-known Harrington desirability scale, as a percentage [28, 29]. The following recommendations are proposed for making decisions on the level of process quality based on a comprehensive assessment (Table 6).

Table 4 Example of processing an expert opinion sheet

		Indicators i										
Indicators j		1	2	3	4	5	$G_j(1)$	$\gamma_j(1)$	$G_j(2)$	$\gamma_j(2)$	$G_j(3)$	$\gamma_j(3)$
1	Fixing the measuring instrument to the parts	1	2	1	2	1	7	0.28	27	0.20	162	0.30
2	Level of automation	0	1	1	1	0	3	0.12	27	0.20	81	0.15
3	The complexity of control	1	1	1	1	2	6	0.24	27	0.20	108	0.20
4	Qualification of the controller	0	1	1	1	1	4	0.16	27	0.20	81	0.15
5	The complexity of preparation for control	1	2	0	1	1	5	0.20	27	0.20	108	0,20
Sum							25	1	135	1	540	1

Table 5 The results of expert evaluation of the weights of the quality indicators of the process of control of complex parts with the results of expert evaluation

Indicator number	Single quality indicator	Weight γ_i
1	Fixing the measuring instrument to the parts	0.30
2	Level of automation	0.15
3	The complexity of control	0.20
4	Qualification of the controller	0.15
5	The complexity of preparation for control	0.20

5 Conclusions

The research is aimed at improving the process of control of the parameters of complex parts in production.

The research of the concept of complex part is carried out in the work. The classification of complex part surfaces was studied and the analysis of design and technological characteristics was carried out. It is determined that the surface of a complex part can be represented as a complex of elementary surfaces.

The shortcomings of the control process have been studied. Therefore, as a result of the work, a system of quality indicators and a method for their evaluation in the process of control of complex parts were proposed. To this end, the paper offers recommendations for decision-making regarding the quality level of the control process. The application of the proposed algorithm makes it possible to quantify the quality of the process of control of complex parts, as well as to analyze in detail the potential inconsistencies.

Table 6 Recommendations for making decisions on the quality of the process of control of complex parts

Quality level, %	Recommendations for decision making
Above 80–100	Very good quality. The process meets all requirements; quality is guaranteed.
Above 63–80	Good quality. Minor discrepancy. The process may be acceptable provided corrective action is taken; there are several minor discrepancies (or one significant one)
Above 37–63	Satisfactory quality. Significant discrepancy. The process is satisfactory. It can be credited conditionally. There are significant inconsistencies that need to be addressed. The process requires the development of a non-compliance analysis procedure and a plan for their elimination (or the development of a corrective action plan)
Above 20–37	Bad quality. Critical compliance of the process. The process requires a time limit to resolve inconsistencies.
Above 0–20	Very poor quality. Complete non-compliance, in which the process requires the development of a corrective action plan, time limits to eliminate non-conformities and re-verification.

The application of the proposed method of quality assessment allows for an objective assessment of the quality level, to make informed decisions in managing the quality of the control process of complex details, as well as to identify inconsistencies (critical points) and criteria for developing a corrective action plan.

In the future, the authors propose to improve the regulatory framework of the control process by developing a regulatory document that will regulate the requirements for quality assessment based on the proposed method and nomenclature of indicators for assessing the quality of the control process of complex details.

Acknowledgments The scientific results have been partially obtained within the research project "Fulfillment of tasks of the perspective plan of development of a scientific direction "Technical sciences" Sumy State University" ordered by the Ministry of Education and Science of Ukraine (State Reg. No. 0121U112684). The research was partially supported by the Research and Educational Center for Industrial Engineering (Sumy State University) and International Association for Technological Development and Innovations. The studies were realized with support from statutory activity financed by the Polish Ministry of Education and Science (0613/SBAD/4677).

References

1. Behun, M., Kascak, P., Hrabcak, M., Behunova, A., Knapcikova, L., & Sofranko, M. (2020). Investigation of sustainable geopolymer composite using automatic identification technology. *Sustainability, 12*, 6377. https://doi.org/10.3390/su12166377
2. Rudenko, S., Berladir, K., Trojanowska, J., Varenyk, S., Shvetsov, D., & Kravets, V. (2021). Application of FMEA for assessment of the polymer composite materials quality. *Journal of Engineering Sciences, 8*(2), B12–B18. https://doi.org/10.21272/jes.2021.8(2).b3
3. Bodein, Y., Rose, B., & Caillaud, E. (2014). Explicit reference modeling methodology in parametric CAD system. *Computers in Industry, 65*(1), 136–147.

4. Salehi, V., & McMahon, C. (2011). Development and application of an integrated approach for parametric associative CAD design in an industrial context. *Computer-Aided Design and Applications, 8*(2), 225–236.
5. Kumar, P., Jain, N. K., & Sawant, M. S. (2020). Modeling of dimensions and investigations on geometrical deviations of metallic components manufactured by μ-plasma transferred arc additive manufacturing process. *The International Journal of Advanced Manufacturing Technology, 107*, 3155–3168. https://doi.org/10.1007/s00170-020-05218-9
6. Savio, E., De Chiffre, L., & Schmitt, R. (2007). Metrology of freeform shaped parts. *CIRP Annals, 56*, 810–835.
7. Li, H., Lu, Q., & Lachmayer, R. A. (2022). Modeling framework to support the implementation of engineering changes in designing complex products. *Research in Engineering Design, 33*, 249–271. https://doi.org/10.1007/s00163-022-00388-x
8. Cusumano, M. A., & Nobeoka, K. (1992). Strategy, structure and performance in product development: Observations from the auto industry. *Research Policy, 21*(3), 265–293.
9. Hovorun, T. P., Berladir, K. V., Pererva, V. I., Rudenko, S. G., & Martynov, A. I. (2017). Modern materials for automotive industry. *Journal of Engineering Sciences, 4*(2), F8–F18. https://doi.org/10.21272/jes.2017.4(2).f8
10. Dehtiarov, I. (2017). *Technological support of processing of details of type of levers in the conditions of serial production with use of flexible machine tools*. Ph.D. thesis, 200 [In Ukrainian].
11. Ivanov, V., Dehtiarov, I., Zaloga, V., Kosov, I., & Savchuk, V. (2020). Increasing productivity of connecting rods machining. In V. Ivanov, J. Trojanowska, I. Pavlenko, J. Zajac, & D. Peraković (Eds.), *Advances in design, simulation and manufacturing III. DSMIE 2020* (Lecture Notes in Mechanical Engineering). Springer. https://doi.org/10.1007/978-3-030-50794-7_26
12. ISO 9001:2015 – Quality management systems — Requirements.
13. Zaloga, V., Yashyna, T., & Dynnyk, O. (2018). Analysis of the theories for assessment of the quality management product efficiency. *Journal of Engineering Sciences, 5*(2), B1–B6. https://doi.org/10.21272/jes.2018.5(2).b1
14. Behúnová, A., Knapčíková, L., & Behún, M. (2020). Logistics of controlling implementation in conditions of manufacturing enterprise. *Acta Logistica, 7*(1), 23–29. https://doi.org/10.22306/al.v7i1.154
15. Lasinska, N. (2021). Hybrid management methodology for transport projects related to rolling stock. *Journal of Engineering Sciences, 8*(2), B7–B11. https://doi.org/10.21272/jes.2021.8(2).b2
16. Ivanov, V., Dehtiarov, I., Pavlenko, I., Kosov, I., & Kosov, M. (2019). Technology for complex parts machining in multiproduct manufacturing. *Management and Production Engineering Review, 10*(2), 25–36.
17. OKRB 500-94. (1994). *Design and technological classifier of products of mechanical engineering and instrumentation*. 71.
18. Ning, F., Shi, Y., Cai, M., & Xu, W. (2020). Various realization methods of machine-part classification based on deep learning. *Journal of Intelligent Manufacturing, 31*(8), 2019–2032. https://doi.org/10.1007/s10845-020-01550-9
19. Klimenko, L. P., Pisintsali L. V., Alexandrovskaya, N. I., & Evdokimov, V. D. (2011). *Metrology, standardization and quality management*.
20. Guowei, Y., & Yizhong, W. (2022). Three-dimensional measurement of precise shaft parts based on line structured light and deep learning. *Measurement, 191*, 110837. https://doi.org/10.1016/j.measurement.2022.110837
21. Sun, Q., Hou, Y., Tan, Q., & Li, C. (2014). Shaft diameter measurement using a digital image. *Optics and Lasers in Engineering, 55*, 183–188.
22. Ainsworth, I., Ristic, M., & Brujic, D. (2000). CAD-based measurement path planning for free-form shapes using contact probes. *The International Journal of Advanced Manufacturing Technology, 16*, 23–31.
23. Leach, R. K., Bourell, D., Carmignato, S., Donmez, A., Senin, N., & Dewulf, W. (2019). Geometrical metrology for metal additive manufacturing. *CIRP Annals, 68*, 677–700.

24. Kysylevska, A. Y., & Nikipelova, O. M. (2012). Qualimetric assessment of the quality of the organization of the release of procedures using therapeutic waters. *Journal of Physiotherapy and Curothology, 2*, 25–26.
25. Denysenko, Y. O., Zaloga, V. O., & Ivchenko, O. V. (2020). *Improving the normative base of instrumental preparation of production for the improvement of technical and economic indicators: A monograph*. SumDU.
26. Azgaldov, G. G. (1982). Theory and practice of assessing the quality of goods. Basics of qualimetry. *Economics*.
27. Azgaldov, G. G. (1977). *Assessment and certification of quality in construction*. Stroyizdat.
28. Denysenko, Y., Ivanov, V., & Ivchenko, O. (2017). Quality assessment of teaching the disciplines in the e-learning environment of Sumy State University. *CEUR Workshop Proceedings, 1844*, 166–175.
29. Harrington, E. C. (1965). The desirability function. Industrial quality. *Control, 21*(10), 494–498.

Fundamentals of Monitoring the Technological Parameters by Deep Drawing Process: A Study

Matúš Martiček, Rebeka Tauberová, Lucia Knapčíková, and Jozef Husár

1 The State of the Art

Drawing is a technological process in which there is a permanent transformation of a flat starting semi-finished product (sheet metal) into a three-dimensional hollow body. The final product of which is moulding, resp. (Fig. 1). The transformation of a planar blank into a hollow body is realized by a forming tool (drawing tool) on a forming machine (mechanical or hydraulic press) [1, 2].

Drawing is a highly productive technological process that belongs to the surface forming group. It achieves the desired shape of the moulding (yield). It is an integral part of industrial production [1]. It is possible to produce hollow thin-walled, sufficiently rigid, and complex low-weight components by drawing. Pulling makes components used in all areas of our lives. All body parts of cars and trucks are produced by pulling [2] in the automotive industry. In households, all-metal kitchen utensils, refrigerators, washing machines, ovens, etc., are made by removing [3]. Towing as a technological operation is also widely used in the manufacture of aircraft, ships, toys, the food industry, and electrical engineering [4]. Material and tool costs account for the largest share of pulling costs. The drawn sheet must be held in most cases when pulling, therefore, the machine must have two slides – pulling and holding [1]. The principle of drawing is that the semi-finished product (blank) is placed on the upper edge of the die, and by moving the die, the sheet metal is pulled into its hole. Any corrugation of the edges of the blank shall be prevented as necessary by the retainers [5, 6]. The bottom of the extract is formed on the face

M. Martiček · R. Tauberová · L. Knapčíková (✉) · J. Husár
The Technical University of Košice, Faculty of Manufacturing Technologies with a seat in Prešov, Department of Industrial Engineering and Informatics, Prešov, Slovak Republic
e-mail: matus.marticek@student.tuke.sk; rebeka.tauberova@tuke.sk; lucia.knapcikova@tuke.sk; jozef.husar@tuke.sk

L. Knapčíková, D. Peraković (eds.), *7th EAI International Conference on Management of Manufacturing Systems*, EAI/Springer Innovations in Communication and Computing, https://doi.org/10.1007/978-3-031-22719-6_13

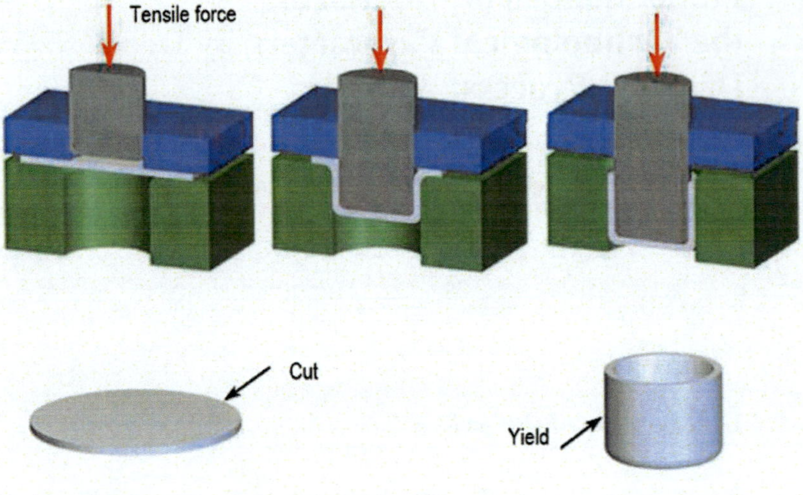

Fig. 1 Deep drawing of the cylindrical yield [2]

of the die, and the blank is bent against its direction of movement at its rounded edges. Depending on the size of the bending radius and the type of material, the sheet thickness in these places will decrease by 10–20%, in rare cases, even more than 20%. The bottom bend is spatial [4]. By further moving the die, the material is compressed in the tangential direction in the entire outer part to pass through the hole in the die. The yield elongates in this radial direction, and the wall thickness of the yield increases by 20–30%. The increase in wall thickness and elongation depends on the degree of drawing. The bottom material is slightly elongated (1–2%) and thins (approximately 3%) [7].

1.1 Influence of Factors on the Process of Deep Drawing

In any cold working of sheets and strips, one of the most important properties is its ability to deform plastically (formability). Formability is defined as overcoming permanent change without disturbing the body (yield) in specific and conditioned technological conditions [8]. Although considerable efforts have been made so far to find only one universal value that would characterize the suitability of the material for different technological forming operations, this goal has not been achieved. Metal sheet and strip processors use various test methods that depend on the specific requirements [9]. The formability of sheets is most often considered during drawing because it is the most complex technological operation in terms of material stress and technical-technological conditions of its implementation. Drawing is a complex process of transforming a flat sheet into a hollow, three-dimensional, often

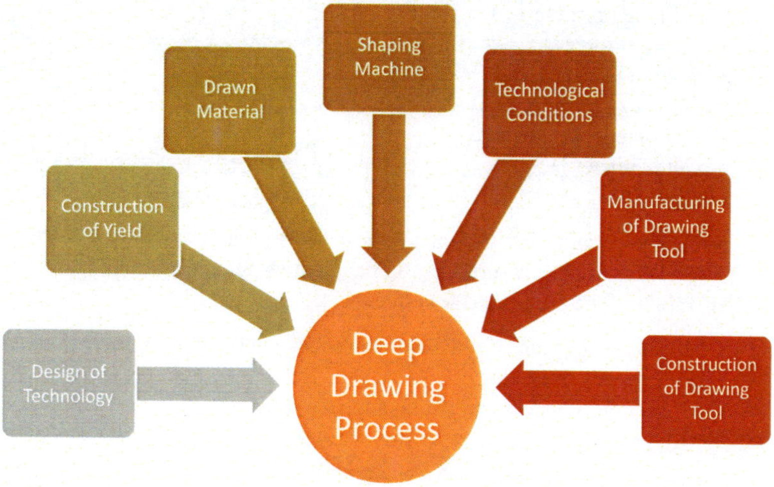

Fig. 2 Influence of factors on the process of deep drawing. (Authors own processing)

complex body (Fig. 2). The outcome of this operation is affected by all the factors involved in it [4, 10]. The condition for successful drawing is the product of the required parameters, shape, without undulations, local thinning, sufficient rigidity, and dimensional stability. It is known from the experience of presses that even though the drawing process has been going on for a long time (on the same machine and tool), there are problems with the yield parameters at certain times [11].

The deep-drawing process converts a flat blank into a three-dimensional hollow, symmetrical or asymmetrical body using drawing tools and forming tools. Drawing tools have an individual character, i.e. for each shape of the moulding, material thickness, drawing method, etc., a particular towing tool is required [12]. The individuality of the deep drawing process places high demands on their design process, especially in the case of drawing complex, asymmetrical yield shapes. In addition to the above-mentioned deep-drawing process, it can be considered a complex multifactor system [13]. The extraction of an intact yield is affected not only by the material's formability but also by friction conditions holding forces, deformation rate, and the like [14].

2 Methodology of Drawing Conditions on Technological Characteristics of the Deep Drawing Process

The conditions under which the pulling process takes place on the press greatly affect the result. According to current knowledge, the technological conditions that most affect the drawing process include:

- Holding pressure
- Traction force
- Bottom tearing force
- Degree of pulling
- Ultimate towing degree
- Cut diameter [15]

The role of the retainer in the pulling process is to prevent the formation of waves in the flange area. For the retainer to perform this function, the material must be subjected to such pressure that the flange does not corrugate [11, 16]. The radii of curvature of the traction edges affect [17]:

- The amount of tension in the drawn sheet and the pulling force
- Limit levels of pulling
- Formation of waves and shifts in the yield

2.1 Friction Processes

Friction is the resistance to the relative movement between two bodies pressed together in the contact area of their surfaces in the tangential direction. This definition does not fully describe internal friction but only external friction. The contact of two friction bodies characterizes external friction. Internal friction occurs in material layers belonging to the same friction body [18].

Effects of external friction in the forming process:

- Affects the condition, course and uniformity of deformation, as well as the mechanical properties of the formed metal
- Has a primary impact on wear and thus on the service life of machines and tools
- Affects the stress and strain in the formed volume and the degree of strain in the vicinity of the contact surfaces
- Affects the distribution, the size of the deformation resistance and affects the resulting force effect in the forming tool
- Affects the energy consumption required in the forming process
- Affects the entire course of the forming process

From the tribological system, we can determine four basic states of friction:

1. Friction of solids occurs when the determining material area is in the solid state. This friction state is further divided into pure friction of the solids, the so-called deformation friction (which occurs mainly in a vacuum) and friction in adhesive layers.
2. Liquid friction is characterized by a state in which the layer of material in which the friction takes place has the properties of a liquid.
3. Gas friction is similar to liquid friction, except the characteristic layer has gas properties.

4. Plasma friction is a condition where the characteristic layer in which the friction process takes has plasma properties [18].

Depending on whether there is a lubricant, another substance or the contact surface of the contact pairs without an intermediate layer, the friction is divided between the contact surfaces of the tribological pairs [18]:

- Dry friction
- Limit friction
- Liquid friction
- Mixed friction [18]

Dry friction – has two aspects: deformation, in which mechanics is more applied, and adhesives, in which chemistry is more used. Both are affected by the properties of the surfaces, their size, thermal properties, mechanical properties, and chemical composition, including the impurities present, the adsorbed, reactive and deposited solid layers, and the chemical reactivity. The presence of dry friction is undesirable in the manufacture of mouldings because the material is seized [18, 19].

Limit friction – if there is already a thin layer of adsorbed molecules of gas, liquid or substance formed by the chemical reaction of the surfaces between the friction surfaces, we speak of boundary friction. In technical practice, limited friction arises in cases where the friction surfaces are loaded by high pressures at small contact surfaces and low shear rates [18, 19].

Liquid friction – creates a continuous layer of lubricant between the friction surfaces, which completely balances the roughness of the surfaces and, with its pressure, cancels the effect of perpendicular loading on the surfaces of the bodies. Thus, the friction force depends only on the amount of internal friction in the lubricating layer in fluid friction conditions. If the lubricating layer is to withstand the loads by which the bodies are pressed together, a corresponding counterpressure must be created in it. This backpressure can be achieved hydrostatically and hydrodynamically [18, 19].

Mixed friction – represents the most widespread form of friction in the production of mouldings. This type of friction occurs when a layer of lubricant does not perfectly separate the friction surfaces, and their unevenness comes into direct contact, i.e., both liquid and boundary friction occur. The proportion of these two components greatly affects the final quality of the moulding. The ratio of liquid and limiting friction largely depends on the roughness of the surfaces. The tool generally has a lower surface roughness than sheet metal [18, 19].

Contact or contact between active members of the tribological system is the primary feature and phenomenon of the whole system when performing the required technical function. Each pressing process has a different type of contact between the sheet metal and the contact surfaces of the drawing tool. As already mentioned, mixed friction most often occurs during the press operations between the contact surfaces of the tool and the sheet metal blank [20].

The effects of mixed friction in the relative motion of the friction pairs cause the release of material particles from the surface. The protrusions themselves (micro-irregularities) have different conditions. The sizes and distances between the protrusions before and after deformation are different. As the friction pairs move relative to each other, the state of the surface microgeometry is constantly changing. The friction pair contact model can be compiled by measuring changes in these areas and obtaining a larger statistically significant set of these data for different operations. Similarly, the microgeometry of the sheet surface has an effect on compressibility [21].

To evaluate the quality of the machined surface, it is sufficient to enter the roughness expressed by some mean value of the height of the unevenness. Additional parameters must be specified to determine friction, as the force of the pressing force does not deform equally sharp protrusions and blunt protrusions [22]. Micro-roughness sensors record profile curves in parallel sections perpendicular to the surface [23]. From such records, we can deduce the size and shape of micro-irregularities. For measuring and recording micro-roughness, mechanical roughness gauges are used as profilers, where the movement of the predominantly diamond tip is sensed along the surface of the surface, or optical roughness gauges are used for the principle of visual reflection and interference is used [24]. This principle is implemented by the motor movement of the sensor over the tested surface.

3 Results and Discussion

Various parameters enter the tribological system in the towing process, which needs to be examined and evaluated. These results can be used in software for forming simulations. This process is most affected by:

- Microgeometry of the sheet surface (surface texture and roughness of this texture)
- Sheet metal feed speeds
- Material deformation (sheet metal switching, deep drawing)
- Mechanical properties of the material and chemical composition
- The amount of contact pressure between the material and the tool
- Type of sheet metal used (surface treatment)
- The type of tool material and the condition of its surface [19, 24, 25]

For this reason, when examining the friction conditions in the tool-lubricant-material-technological tribological system, test results are often used, which model the situations in the individual areas of contact between the tool and the sheet metal. Therefore, friction simulators are proposed that simulate the stress conditions in the particular contact areas of the sheet metal and the tool in the pressing process [25].

The radial sheet simulator, which simulates friction at the drawing edge (Fig. 3), consists of a tensioning mechanism using which the braking force is derived. The tensioning tool is based on the braking rib principle [24]. The fixed rollers are fixed, and the movable rollers can be adjusted with screws.

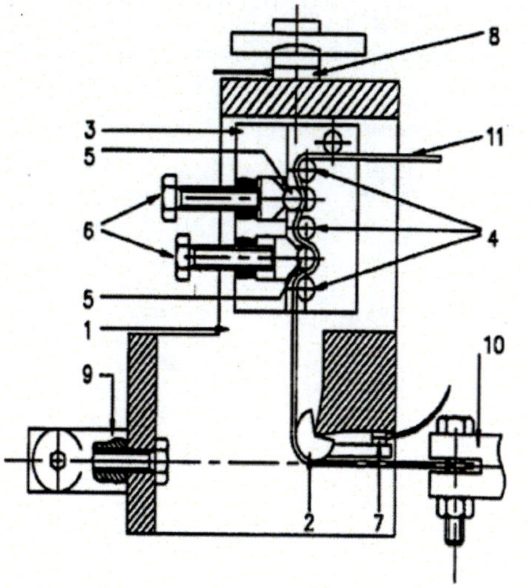

Fig. 3 Radial traction simulator [2]. Simulator description [25]: (1) cover, (2) radius, (3) tensioning mechanism, (4) fixed rollers, (5) movable rollers, (6) screws, (7) and (8) piezo elements, (9) fixed head, (10) movable head, (11) stripe

The morphology of the sheet surface also greatly influences the development of deformation and friction in the tribological system [26]. Characteristics such as roughness, shape and number of peaks affect the friction ratios on the contact surfaces of the tool and the sheet during pressing. The surface microgeometry of the material is also important in appearance and pressure adhesion. The sharp tip of the profilometer sensor converts the distribution of the unevenness on the surface into a mechanical movement, which is then converted by the sensor into an electrical signal and further interpreted as a numerical value of the selected surface structure parameter, or as a visual record of the surface roughness profile [27].

Wear is an undesirable change in the surface or dimensions of the solids, caused either by the interaction of the functional surfaces or by the functional surface and the medium which causes the wear. It is manifested by the removal or displacement of matter particles from the functional surface by mechanical effects, sometimes accompanied by other influences, such as chemical or electrochemical [28].

The wear process has a characteristic course that can be divided into three parts (Fig. 4).

I. Running-in, at this stage at the beginning the wear is high. Some micro-irregularities are removed, and balanced surface roughness is achieved. The contact area changes and broken particles also have a significant effect. Gradually, the surfaces become more polished, and the resulting unevenness gradually disappears, and the wear decreases. After a certain time, there is no progressive increase in wear. The wear products are eliminated. Run-in is affected by the number of mechanical wear processes.
II. Normal course, operation – characterizes a linear process, where the wear increases linearly with time. The aim is to achieve as little wear and tear as possible at this stage. After a certain period of operation, the wear has

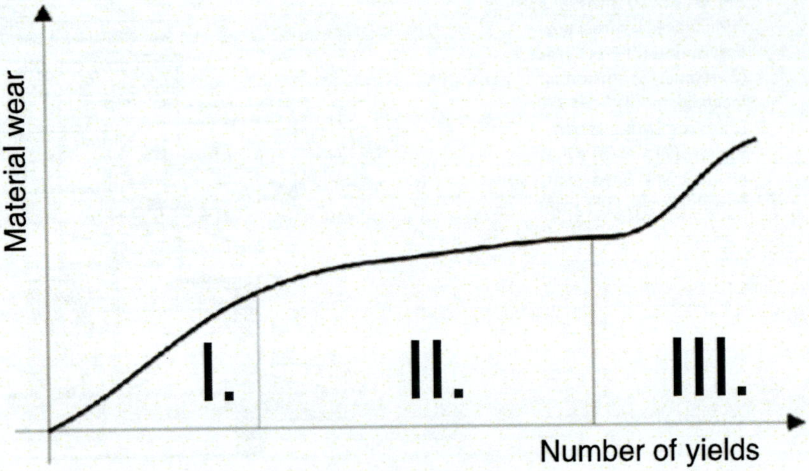

Fig. 4 Basic phases of material wear. (Authors own processing)

increased, and the wear begins to rise progressively. It can be caused by the gradual removal of the surface layer, and the accumulation of minor damage (fatigue). Large material removal can be caused by larger, broken, highly abrasive particles.

III. Emergency – is the final stage of wear and tear, the so-called life expectancy. Wear that occurs during shear friction is divided into wear by seizing, wear by abrasion (loss of material), and wear by shape change. Most attention is paid to abrasion wear, resulting in undesirable material loss [28].

Total wear is primarily affected by the length of the production run and the complexity of the forming operation. This complete wear can be produced by abrasion, adhesion, or both. The wear can vary considerably, depending on: – the surface characteristics of the sheet – the technological elements of the deep drawing – the lubricant used – the nature of the load – the type of drawing process (without a retainer, with retainer) [28]. There are two types of wear on the functional parts of the drawing tool, namely: – even wiping until the prescribed dimensions are lost – gluing the material from the blank to the working part of the drawing tool [27, 29].

During deep drawing, the material of the extract is moved, and for this reason, the material is glued to the functional surfaces of the drawing tool. The adhesion of particles of material to the die and its peeling also changes the nature of the friction between the die and the die from adhesive to abrasive [30]. This is then manifested by the wear of the functional surfaces in the form of scratches and depressions. Due to wear, the dimensions, shape and surface quality of the available parts of the tool and thus the quality of the yields gradually change. The difference between the friction ratios at the radius of curvature of the block and the line is that the slip speed on the block is significantly lower than at the edge of the block. The ratios at the radius of curvature of the line can be described as dynamic, while on the beam, they are more static than dynamic [31].

4 Conclusion

Today, there are increasing demands on mouldings' quality and technical and economic levels. For this reason, it is necessary to determine as precisely as possible some of the pressing characteristics which would help to ensure the most accurate and reliable data possible in the processes of simulation and prediction of sheet metal capability. A suitable material and friction models must be available to match the simulation results with the test results. Frictional forces on the contact surfaces of the tool with the sheet affect the technological characteristics of the process and the resulting quality of the stamping. The coefficient of friction can express friction ratios, the value of which must be entered when defining the input data. Coating tool coating is used to reduce friction, seizure, and wear. At the same time, the base material must be sufficiently hard and strong so that it is not deformed under load. Otherwise, the coating would also break. Emphasis is mainly placed on the layers used in the forming processes for their sufficient wear resistance, hardness, adhesion, and sufficient heat resistance. Coating of traction tools is divided into deposit and reactive. In reactive coating, the thin or chemical-thermally modified thin surface layer of the material is alloying elements. The future direction of the research will be focused on the online monitoring of technological parameters by the deep drawing process. The future direction of research will be focused on selected tools of Industry 4.0, as well as a digital twin and the Internet of Things.

Acknowledgements This work was supported by the project VEGA 1/0268/22-granted by the Ministry of Education, Science, Research and Sport of the Slovak Republic.

References

1. Technology of surface forming-drawing. *Technology II*. TU Liberec. http://www.ksp.tul.cz/cz/kpt/obsah/vyuka/skripta_tkp/sekce/09.htm. Accessed 22 February 2022.
2. Marticek, M. (2010). *Study of friction conditions during the deep drawing of cylindrical extracts*. Master thesis, TU Košice.
3. Mold Technology. *Deep drawing*. http://mold-technology4all.blogspot.cz/2011/08/deep-drawing.html. Accessed 15 March 2022.
4. Fichema. http://www.fichema.cz/. Accessed 6 March 2022.
5. Get Kaung. *Metal deep drawing of car panel*. http://www.getkaung.com.tw/mobile/metal-deepdrawing/metal-deep-drawingof-car-panel.html. Accessed 21 April 2022.
6. Quia. *Conventional spinning*. https://www.quia.com/jg/1961543list.html
7. *Metal deep drawing process of sheet metal forming*. http://techminy.com/metal-deep-drawingprocess. Accessed 20 April 2022.

8. ITWDrawform. *Product features*. http://www.itwdrawnmetals.com/product-features/default.html. Accessed 12 March 2022.
9. Pacaiova, H., Sinay, J., Turisova, R., et al. (2017). Measuring the qualitative factors on copper wire surface. *Measurement, 109*, 359–365.
10. Wang, L., Zhou, H., & Li, N. (2020). Carrier transport composites with suppressed glass-transition for stable planar perovskite solar cells. *Journal of Materials Chemistry A, 8*(28), 14106–14113.
11. Heath, G., Silvermann, T., Kempe, M., et al. (2020). Research and development priorities for silicon photovoltaic module recycling to support a circular economy. *Nature Energy, 5*(7), 502–510.
12. Straka, M., Khouri, S., et al. (2018). Utilization of computer simulation for waste separation design as a logistics system. *International Journal of Simulation Modelling, 17*(4), 83–596.
13. Knapčíková, L., Herzog, M., & Oravec, P. (2010). Material characterization of composite materials from used tires. *Výrobné inžinierstvo, 4*, 31–34.
14. Liu, J., Gaynor, A. T., Chen, S., Kang, Z., Suresh, K., Takezawa, A., Li, L., Kato, J., Tang, J., Wang, C. C. L., et al. (2018). Current and future trends in topology optimization for additive manufacturing. *Structural and Multidisciplinary Optimization, 2018*(57), 2457–2483.
15. Straka, M., Rosova, A., et al. (2018). Principles of computer simulation designfor the needs of improvement of the raw materials combined transport system. *Acta Montanistica Slovaca, 23*, 163–174.
16. Kaufmann, S. A. (1993). *The origins of order: Self-organization and selection in evolution*. Oxford University Press.
17. Mandičák, T., Mesároš, P., & Tkáč, M. (2018). Impact of management decisions based on managerial competencies and skills developed trough BIM technology on performance of construction enterprises. *Pollack Periodica, 13*(3), 131–140.
18. Hosford, W. F. (2010). *Mechanical behavior of materials*. Cambridge University Press.
19. Fiebig, S., Sellschopp, J., Manz, H., Vietor, T., Axmann, K., & Schumacher, A. (2015). Future challenges for topology optimization for the usage in automotive lightweight design technologies. In *Proceedings of the 11th world congress on structural and multidisciplinary optimization*, Sydney, Australia, 7–12 June 2015, Vol. 142, pp. 1–8.
20. Singh, R., & Davim, J. P. (2018). *Additive manufacturing: Applications and innovations* (1st ed., p. 280). CRC Press.
21. Li, N., Huang, S., Zhang, G., Qin, R., Liu, W., Xiong, H., Shi, G., & Blackburn, J. (2019). Progress in additive manufacturing on new materials: A review. *Journal of Materials Science and Technology, 2019*(35), 242–269.
22. Sinay, J., Balazikova, M., Dulebova, M., et al. (2018). Measurement of low-frequency noise during CNC machining and its assessment. *Measurement, 119*, 190–195.
23. Liu, J., Gaynor, A. T., Chen, S., Kang, Z., Suresh, K., Takezawa, A., Li, L., Kato, J., Tang, J., Wang, C. C. L., et al. (2018). Current and future trends in topology optimization for additive manufacturing. *Structural and Multidisciplinary Optimization, 57*, 2457–2483.
24. Kanáliková, A., Behúnová, A., Mandičák, T., & Mesároš, P. (2019). New technologies for sustaining development in infrastructure, logistics and construction industry. *Acta Logistica, 6*(4), 171–177.
25. Al-Salamah, M. (2016). Economic production quantity in batch manufacturing with imperfect quality, imperfect inspecyion and non-desructive acceptance sampling in a two-tier market. *Computers and Industrial Engineering, 93*, 275–285.
26. Lin, C. C., & Wang, T. H. (2011). Build-to-order supply chain network design under supply and demand uncertainties. *Transportation Research Part B, 45*, 1162–1176.
27. Smeringaiova, A., & Wittner, M. (2018). Parametric modeling as efficient way of designing. *MM Science Journal, 12*, 2701–2705.
28. Gaspar, S., & Pasko, J. (2018). Mathematical formulation of the kinematic equations for the control of the robot system with application for the machining conical surfaces. *MM Science Journal, 2018*, 2158–2161.

29. Merzliakov, I., Pavlenko, I., Chekh, O., Sharapov, S., & Ivanov, V. (2020). Mathematical modeling of operating process and technological features for designing the vortex type liquid-vapor jet apparatus. In V. Ivanov et al. (Eds.), *Advances in design, simulation and manufacturing II. DSMIE 2019* (Lecture Notes in Mechanical Engineering) (pp. 613–622). Springer. https://doi.org/10.1007/978-3-030-22365-6_61
30. Voloshina, A., Panchenko, A., Panchenko, I., Titova, O., & Zasiadko, A. (2019). Improving the output characteristics of planetary hydraulic machines. *IOP Conference Series: Materials Science and Engineering, 708*(1), 012038. https://doi.org/10.1088/1757-899X/708/1/012038
31. Khovanskyi, S., Pavlenko, I., Pitel, J., Mizakova, J., Ochowiak, M., & Grechka, I. (2019). Solving the coupled aerodynamic and thermal problem for modeling the air distribution devices with perforated plates. *Energies, 12*(18), 3488. https://doi.org/10.3390/en12183488

A Heuristic Project Buffer Management Algorithm in the Scheduling Phase of a Renewable Energies Project

Shakib Zohrehvandi and Lucia Knapčíková

1 Introduction

Today, one of the biggest problems for companies and organizations in the implementation process is the failure to complete projects according to schedule. Project planning is a complex process that involves various resources and activities that need to be optimized [1]. The resource-constrained project scheduling problem (RCPSP) is one of the well-known problems that should be considered in project activities to minimize project duration. Due to the nonoptimality of the schedule and not completing it on time, it is often possible to use the optimization method using the critical chain project management (CCPM) method.

One of the optimization methods to increase the stability of the project schedule is to create a buffer (safety time) to deal with time changes in the project using the critical chain method. The buffer of the project is added at the end of the critical path, which does not float so that the desired buffer can be used in case of delays in the critical path. Types of buffers include project buffer (PB), feeding buffer (FB), and resource buffer (RB). The project buffer is placed at the end of the project life chain to maintain the project delivery date. A feeding buffer is added to the paths that connect to the critical chain so that possible delays do not affect the critical chain.

S. Zohrehvandi (✉)
Faculty of Manufacturing Technologies with a seat in Prešov, Department of Industrial Engineering and Informatics, The Technical University of Košice, Prešov, Slovak Republic

New Technologies Department, Center for European Studies, Kharazmi University, Tehran, Iran
e-mail: shakib.zohrehvandi@srbiau.ac.ir

L. Knapčíková
New Technologies Department, Center for European Studies, Kharazmi University, Tehran, Iran
e-mail: lucia.knapcikova@tuke.sk

L. Knapčíková, D. Peraković (eds.), *7th EAI International Conference on Management of Manufacturing Systems*, EAI/Springer Innovations in Communication and Computing, https://doi.org/10.1007/978-3-031-22719-6_14

Resource buffers ensure that project resources are ready when a critical activity needs the resource [2].

In 2015, Hall conducted extensive research on project management research opportunities for the next 10 years. In this research, one of the areas of project management that were considered with high potential for future research was the area of project buffer management and project scheduling. One of the most important steps in project buffer management is to determine the size of the buffer. The things that should be considered when determining the size of the buffer are the type of project activities, the volume and size of the project activities, the complexities and relationships between the activities, the number of available resources for the activities and the required resources, and the number of uncertainties in the project environment [3]. To create more accurate buffer measurements in this research, the resource availability coefficient, resource flexibility, and resource activity quality have been investigated. The proposed algorithm will be implemented as a case study in the planning phase of a renewable wind farm project.

The purpose of this study is to introduce the HPBM algorithm to determine the project buffer size in the planning phase of a wind farm. The research question is, what is the difference between using the HPBM algorithm in determining the size of the project buffer and determining the size of the buffer using previous similar algorithms?

2 Literature Review

Nowadays, it is very important to develop methods that can be used for the success of projects. The project critical chain management method was first proposed by Goldratt (1997) to improve the traditional project management method by introducing a new mechanism to manage uncertainties [4]. The theory of constraints and its direct application in project management, under the title of critical chain buffer management (CC/BM), is a popular and effective approach in project management, proposed by [5]. Since the publication of Goldratt's theory, several studies have been conducted by the following researchers in this field: [6–12].

Zohrehvandi and Khalilzadeh (2018) conducted a study on project buffer management. In this research, several buffer management methods were first implemented in a wind farm project and the results showed that the adaptive procedure method with resource scarcity (APRT) performs better than other methods. Then the APRT method was combined with the failure mode risk assessment and effects analysis (FMEA) method. After implementing this algorithm, the project was completed in a shorter period than the previous methods [13]. Many researchers have been able to develop and improve traditional project buffer management methods, for example, Bie et al. (2012) introduced a method to determine the buffer size assuming interdependence between activities [14]. Zhang et al. (2017) introduced a method to determine the project buffer size with a fuzzy approach [15]. Sarkar et al. (2018) developed a project life chain management framework for effective implementation in construction projects. The results showed that the proposed framework improved critical chain management by measuring the buffer by integrating various uncertainties that affect the construction schedule [16].

Zohrehvandi et al. (2020) introduced a heuristic algorithm for determining the size of the project buffer and feeding buffers, as well as dynamic buffer consumption control, called the Fuzzy Overlap Buffer Management Algorithm (FOBMA). In this research, pentagonal fuzzy numbers were used to determine the appropriate amount of project activity resources. In addition, an overlay method was used to obtain more realistic durations for the activities. In this research, buffer consumption is controlled using a dynamic method. Considering the different conditions in different phases of a project (i.e., the duration of each phase, the number of resource activities, and the complexity of the network of activities), it is necessary to control the buffer consumption dynamics. For this purpose, the number of buffers that remain unused in each stage of the project is transferred to the next stage [17]. In another study, they introduced an efficient project buffer and resource management (PBRM) model to level project resources and project buffer size and control project buffer consumption of a wind farm project to achieve a more realistic project duration [18]. In addition, they proposed a project time optimization algorithm that first calculates project buffers and feeding buffers, and then dynamically controls buffer consumption in different phases of a wind farm project to find more realistic project durations [19].

Poshdar et al. (2016) discussed a probability-based buffer allocation method in which the buffer size was determined by project planners based on priorities [20]. Bakry et al. (2016) proposed a buffer size algorithm for the optimal planning of construction projects under uncertainty and used fuzzy set theory to model uncertainties associated with input parameters [21]. Zohrehvandi and Soltani (2022) reviewed the latest models and methods of project buffer management and time optimization of construction projects and manufacturing industries [22]. Zohrehvandi (2022) introduced a heuristic algorithm to plan the construction projects of a power plant and project resource management to determine the size of project buffers and feed buffers. This algorithm includes three steps: (1) estimating the duration of project activities, (2) determining the size of the project buffers and feeding buffers, and (3) simulating the mentioned algorithm [23].

Alfieri et al. (2016) proposed an exact method to solve the project buffer allocation problem [24]. Hu et al. (2017) introduced an improved buffer management framework in the project's critical chain, which enabled the allocation of additional resources [25]. Ghoddousi et al. (2017) presented a buffer sizing method aimed at maximizing production schedule efficiency, and at the same time, project risk analysis was conducted to investigate externalities in buffer sizing [26]. Zohrehvandi (2022) examined modeling and optimization in project planning and scheduling in construction management and project time optimization [27]. In another research, he introduced a fuzzy project buffer management (FPBM) algorithm, which is a combination of APRT and fuzzy failure mode and effect analysis (FFMEA) methods [28].

Ghoddousi et al. (2017) proposed a two-stage multi-objective project buffer allocation method for detailed project planning [29]. Martens and Vanhouck (2017) proposed a buffer control method that manages the acquisition value of buffer allocation in different phases of the project [30]. Zhang et al. (2018) introduced a buffer control model based on the characteristics of different phases of the project [31]. A predictive buffer management algorithm (PBMA) based on machine learning technology has been proposed to predict the size of the project buffer and control the consumption of the project buffer in construction projects so that it does not exceed the predetermined schedules and thus avoid delays in project completion time [32].

According to the literature review, some of the previous research on project buffer management has been traditional methods, which were mainly based on limited resources, and their results were not accurate enough.

3 Research Methodology

The HPBM algorithm is shown schematically in Fig. 1. In general, the HPBM algorithm has two steps: preparing the project schedule and determining the size of the project buffer.

According to the algorithm, in the first phase, after determining the appropriate duration of the project activities, the relationships between the activities are determined, and finally, the path and number of critical activities of the project are determined. Then, in the second stage, by holding meetings with experts and

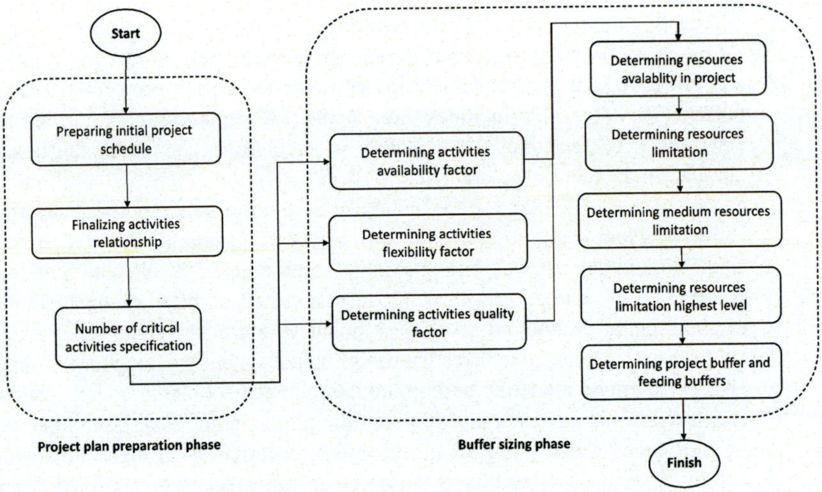

Fig. 1 The heuristic project buffer management algorithm

elites related to the project, the availability factor of the activity resources, the flexibility factor of the activity resources, and also the quality factor of the activity resources are determined to determine the number of project resources. Limits are established. Then the amount of resources available in the project is calculated and then the amount of project resource limitations is determined. At this stage, the average resource limit of each activity and the highest level of resource limit of each activity in the project will be determined. Then the size of the project buffer plan is determined. To evaluate the performance of the proposed innovative algorithm, the algorithm is coded and implemented in MATLAB software. The indicators, parameters, and variables used in the proposed algorithm of this research are as follows:

Indices:
Project activity index: $(u = 1, \ldots, U)$
Project resource index: $(q = 1, \ldots, Q)$
Project critical activity index: $(c = 1, \ldots, C)$

Parameters:
D_u: The duration of project activities
A_q: Activity's resource availability factor
F_q: Activity's resource flexibility factor
Q_q: Activity's resource quality factor
$\mathrm{Rpr}(u, q)$: Availability of resource q for activity u
V_u: The ratio of the duration of activity u in the critical chain
PB^{A}: The total size of the consumed buffer of the project

Decision Variables:
$r(u, q)$: Resource value of type q for activity u
$\mathrm{RC}(u, q)$: Amount of resource constraints q for activity u
RCM_u: Average resource limit activity u
RCH_u: The highest level of resource-constrained activity u
Y: Project buffer compatibility factor
PB^{P}: The planned buffer size of the critical chain of the entire project

In Eq. (1), A_q, F_q, and Q_q are the activity resource availability factor, the activity resource flexibility factor, and the activity resource quality factor, respectively, are very important in determining the extent of RC resource limitations (u, q) and are decisive. In Eq. (1), $\mathrm{Rpr}(u, q)$ is the value of the availability of resource q for activity u.

$$\mathrm{RC}\,(u, q) = \frac{r\,(u, q)}{\mathrm{Rpr}\,(u, q) * Q_q * A_q * F_q} \tag{1}$$

The availability of activities A_q is one of the most important things in resource planning. According to Eqs. (2) and (3), the average activity of the resource limit u and the highest activity level of the resource limit u are calculated respectively. The Q in Eq. (2) represents the number of sources of activity u.

$$\mathrm{RCM}_u = \frac{\sum_{q=1}^{n} \mathrm{RC}\,(u, q)}{Q} \tag{2}$$

$$\mathrm{RCH}_u = \underset{q=1,\ldots,n}{\mathrm{Max}} \;(\mathrm{RC}\,(u, q)) \tag{3}$$

In Eq. (4), Y is the compatibility coefficient of the project buffer, which is obtained as follows:

$$Y = (1 + \mathrm{RCM}_u) * (1 + \mathrm{RCH}_u) \tag{4}$$

V_u also represents the ratio of the duration of activity u in the critical chain. PB^{P} Eq. (5) shows the planned buffer size of the project. In Eq. (5), $u = C$ indicates that the calculated activities are critical.

$$\mathrm{PB}^{\mathrm{P}} = \sqrt{\sum_{u=1}^{U} (Y * V_u)\ \ (u = C)} \tag{5}$$

4 Research Implementation

Figure 2 shows the project schedule for a complete wind turbine, including the number and name of activities, duration, and information about predecessor-successor relationships.

This plan includes the engineering, procurement, and construction phases that are designed in the MSP software. According to the plan, the engineering phase is 75 days, the procurement is 74 days, and the construction phase is 132 days. 162 days have been considered for the implementation of all the mentioned phases, which is the entire implementation of one wind turbine. In this plan, activities 4, 5, 7, 12, 13, 14, 17, 18, 19, and 20 are vital and form the critical path of this project.

5 Results

Through interviews with experts, the number of resources needed to operate in the project stages was prepared, whose values are given in Table 1.

In this step, using the technique presented in the research method section, the size of the project buffer is determined. First, the activity resource availability factor (A_q), the activity resource flexibility factor (F_q), and also the activity resource quality factor (Q_q) were determined simultaneously to determine the extent of project resource limitations, the results of which can be seen in Table 2.

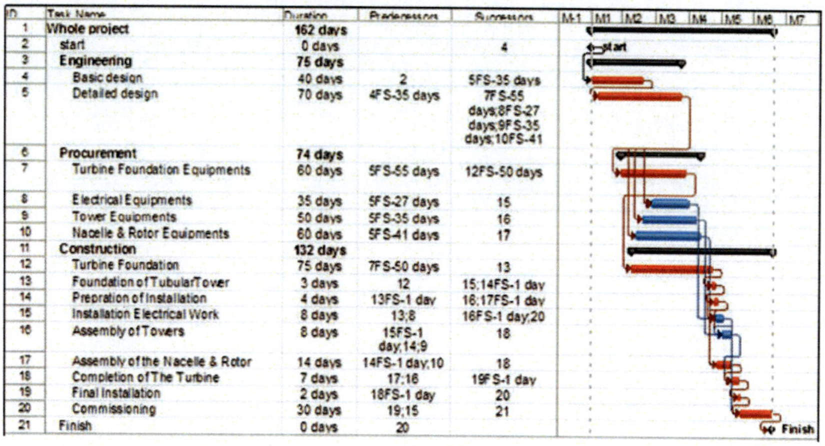

Fig. 2 The project schedule

Table 1 The appropriate amount of project activity resources

Activity number	Activities name	$r(u, q)$				
		$q1$	$q2$	$q3$	$q4$	$q5$
1	The whole project					
2	Project start	–	–	–	–	–
3	Engineering phase					
4	Basic design	5.9	3.8	5.3	–	5.2
5	Accurate and detailed design	6.3	5.3	8.8	5.7	5.9
6	Procurement phase					
7	Foundation equipment	9.9	7.3	8.0	6.1	8.8
8	Electrical equipment	6.2	–	7.4	8.0	5.7
9	Tower equipment	6.5	6.3	5.3	4.9	–
10	Nacelle & Rotor Equipment	9.7	10.8	7.0	10.6	8.1
11	Construction					
12	Turbine base	6.6	5.2	7.8	10.8	8.3
13	Pipe tower foundation	1.9	–	3.4	–	–
14	Installation preparation	4.4	2.6	–	4.7	–
15	Installation of electrical work	5.6	3.7	4.4	–	5.3
16	Assembly of tower	5.9	6.2	–	6.6	7.8
17	Nacelle and rotor assembly	7.5	5.6	5.2	6.3	6.2
18	Turbine completion	5.0	–	6.3	4.4	6.5
19	Final installation	2.8	2.8	4.7	–	4.2
20	Setting up	6.9	–	6.1	6.3	7.6
21	End of the project	–	–	–	–	–

Table 2 Availability, flexibility, and quality factors of activities' resource

	A_q					F_q					Q_q				
Activities name	$q1$	$q2$	$q3$	$q4$	$q5$	$q1$	$q2$	$q3$	$q4$	$q5$	$q1$	$q2$	$q3$	$q4$	$q5$
The whole project															
Project start	–	–	–	–	–	–	–	–	–	–	–	–	–	–	–
Engineering phase															
Basic design															
Accurate and detailed design	0.6	0.4	0.5	–	0.5	0.7	0.5	0.6	–	0.6	0.8	0.6	0.7	–	0.7
Procurement phase	0.6	0.5	0.9	0.6	0.6	0.8	0.6	1.1	0.7	0.7	0.8	0.7	1.0	0.8	0.8
Foundation equipment															
Electrical equipment	1.0	0.7	0.8	0.6	0.9	1.2	0.9	1.0	0.7	1.1	1.0	0.7	0.8	0.6	0.8
Tower equipment	0.6	–	0.7	0.8	0.6	0.7	–	0.9	1.0	0.7	0.6	–	0.7	0.8	0.5
Nacelle & rotor equipment	0.6	0.6	0.5	0.5	–	0.8	0.8	0.6	0.6	–	0.6	0.6	0.5	0.4	–
Construction	1.0	1.0	0.7	1.0	0.8	1.2	1.2	0.8	1.2	1.0	1.0	1.0	0.7	1.0	0.8
Turbine base															
Pipe tower foundation	0.7	0.5	0.8	1.0	0.8	0.8	0.6	0.9	1.2	1.0	0.5	0.4	0.7	0.9	0.7
Installation preparation	0.2	–	0.3	–	–	0.2	–	0.4	–	–	0.4	–	0.6	–	–
Installation of electrical work	0.4	0.3	–	0.5	–	0.5	0.3	–	0.6	–	0.7	0.5	–	0.7	–
Assembly of tower	0.6	0.4	0.4	–	0.5	0.7	0.4	0.5	–	0.6	0.8	0.6	0.7	–	0.8
Nacelle and rotor assembly	0.6	0.6	–	0.7	0.8	0.7	0.7	–	0.8	0.9	0.9	0.9	–	0.9	1.0
Turbine completion	0.7	0.6	0.5	0.6	0.6	0.9	0.7	0.6	0.8	0.7	1.0	0.8	0.8	0.9	0.9
Final installation	0.5	–	0.6	0.4	0.7	0.6	–	0.8	0.5	0.8	0.7	–	0.9	0.7	0.9
Setting up	0.3	0.3	0.5	–	0.4	0.3	0.3	0.6	–	0.5	0.5	0.5	0.7	–	0.7
End of the project	0.7	–	0.6	0.6	0.8	0.8	–	0.7	0.8	0.9	1.0	–	0.9	0.9	1.0
The whole project	–	–	–	–	–	–	–	–	–	–	–	–	–	–	–

Then the resources available in the project (Rpr (u, q)) were calculated and then the amount of project resource limitation (RC (u, q)) was determined, which is given in Table 3.

In this step, the average resource limitation of each activity (RCMu) and the highest level of resource limitation of each activity (RCHu) were determined in the project. The results are shown in Table 4.

The size of the project buffer plan is determined by setting the average resource limit and the highest resource limit level. Finally, the project buffer size (PB^P) was determined to be 21.6 days. The project buffer is located at the end of the critical path of the project shown in Fig. 3.

To evaluate the HPBM algorithm, the inputs of this project were used in several algorithms recently proposed by researchers, and their results are compared in Table 5:

According to Table 5 and based on the evaluations performed, the project buffer measured in this research is 11% lower than the algorithm of Zhang et al. [15], 16% compared to the algorithm of Zohrehvandi and Khalilzadeh [13], and 8% compared to the algorithm of Zhao et al. is less.

6 Disruption and Conclusion

This research presented the HPBM for project buffer management using a modern project buffer management method that was implemented in the planning phase of a renewable wind farm project.

Based on the proposed algorithm, in the first phase, after determining the appropriate duration for the project activities, the relationships between the activities were determined and, finally, the path and number of critical project activities were determined. Then, in the second stage, by holding meetings with experts and elites related to the project, the access factor to the resources of activities, the factor of flexibility of resources of activities, and the factor of quality of resources of activities are simultaneously examined were determined to determine the number of project resources. Then the number of resources available in the project was calculated and then the amount of project resource limitations was determined. At this stage, the average resource limit of each activity and the highest level of resource limit of each activity in the project were determined. Finally, the size of the project buffer plan was determined.

To analyze the data and evaluate the performance of the algorithm, the proposed algorithm was coded and implemented in the MATLAB software, and the inputs from this project were used in several models recently presented by researchers, and their results were compared with each other. According to the research question and according to the results, it was found that the buffer size determined by the HPBM algorithm is more efficient and less than other similar algorithms, which indicates

Table 3 Available resources and amount of project resource constraints

Activity number	Activities name	Rpr (u, q)					RC (u, q)				
		$q1$	$q2$	$q3$	$q4$	$q5$	$q1$	$q2$	$q3$	$q4$	$q5$
1	The whole project										
2	Project start	–	–	–	–	–	–	–	–	–	–
3	Engineering phase										
4	Basic design	7.0	4.0	5.0	–	6.0	2.5	9.5	4.3	–	3.7
5	Accurate and detailed design	5.0	4.0	8.0	7.0	5.0	3.2	5.3	1.2	2.7	3.6
6	Procurement phase										
7	Foundation equipment	9.0	9.0	6.0	5.0	11.0	0.9	1.8	2.3	5.0	1.0
8	Electrical equipment	8.0	–	7.0	6.0	3.0	3.0	–	2.3	2.3	9.6
9	Tower equipment	6.0	8.0	6.0	4.0	–	3.7	2.9	5.5	10.0	–
10	Nacelle & rotor equipment	9.0	8.0	8.0	10.0	6.0	0.9	1.1	2.3	0.9	2.2
11	Construction										
12	Turbine base	5.0	4.0	8.0	9.0	7.0	4.8	10.9	2.0	1.1	2.0
13	Pipe tower foundation	3.0	–	2.0	–	–	36.9	–	21.0	–	–
14	Installation preparation	3.0	3.0	–	4.0	–	9.4	22.6	–	6.3	–
15	Installation of electrical work	5.0	3.0	6.0	–	4.0	3.7	12.3	4.7	–	5.0
16	Assembly of tower	7.0	5.0	–	6.0	6.0	2.4	3.1	–	2.3	1.8
17	Nacelle and rotor assembly	7.0	5.0	7.0	5.0	4.0	1.6	3.7	2.9	3.0	3.8
18	Turbine completion	4.0	–	5.0	3.0	5.0	5.6	–	3.0	9.2	2.8
19	Final installation	4.0	2.0	4.0	–	4.0	14.7	29.5	6.2	–	7.4
20	Setting up	6.0	–	5.0	7.0	5.0	2.0	–	3.1	2.1	2.2
21	End of the project	–	–	–	–	–	–	–	–	–	–

Table 4 The average resource constraint of each activity and the highest level of resource constraint of each activity in the project

Activity number	Activities name	RCM*u*	RCH*u*
1	The whole project		
2	Project start		
3	Engineering phase		
4	Basic design	5.0	9.5
5	Accurate and detailed design	3.2	5.3
6	Procurement phase		
7	Foundation equipment	2.2	5.0
8	Electrical equipment	4.3	9.6
9	Tower equipment	5.5	10.0
10	Nacelle & rotor equipment	1.5	2.3
11	Construction		
12	Turbine base	4.2	10.9
13	Pipe tower foundation	29.0	36.9
14	Installation preparation	12.7	22.6
15	Installation of electrical work	6.4	12.3
16	Assembly of tower	2.4	3.1
17	Nacelle and rotor assembly	3.0	3.8
18	Turbine completion	5.1	9.2
19	Final installation	14.4	29.5
20	Setting up	2.4	3.1
21	End of the project		

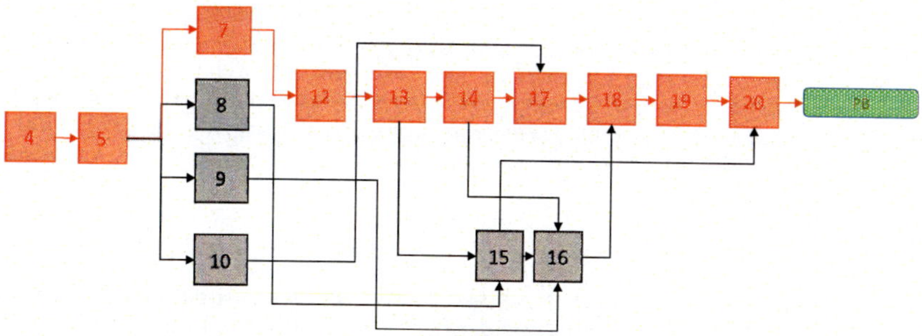

Fig. 3 Location of the project buffer

the better results of this algorithm and other similar algorithms. This research can be useful for researchers, project managers, industrial owners, and all those who work with projects. This research can be implemented in all projects in addition to power plant projects.

Table 5 Comparison of the output of HPBM with the output of previously presented algorithms

Items	Authors	Planned project buffer (days)
1	Zhang et al. [15]	24.2
2	Zohrehvandi and Khalilzadeh [13]	25.8
3	Zhao et al. (2020)	23.5
4	This research (HPBM)	21.6

Acknowledgments This research was funded by the European Union's Horizon 2020 research and innovation program under grant agreement No.723274 and VEGA 1/0268/22 "Design of a Digital Twin for Monitoring the Production Parameters of Technological Equipment Using Augmented Reality".

References

1. Zwikael, O., & Globerson, S. (2004). Evaluating the quality of project planning: A model and field results. *International Journal of Production Research, 42*(8), 1545–1556.
2. Vanhoucke, M. (2016). Buffer management. In *Integrated project management sourcebook* (pp. 155–193). Springer.
3. Hall, N. G. (2015). Further research opportunities in project management. In *Handbook on project management and scheduling* (Vol. 2, pp. 945–970). Springer.
4. Goldratt, E. M. (1997). *Critical chain*. North River Press.
5. Goldratt, E. M., & Cox, J. (1984). *The goal: Excellence in manufacturing*. North River Press.
6. Newbold, R. C. (1998). *Project management in the fast lane: Applying the theory of constraints*. CRC Press.
7. Herroelen, W., & Leus, R. (2001). On the merits and pitfalls of critical chain scheduling. *Journal of Operations Management, 19*(5), 559–577.
8. Leach, L. P. (2005). *Critical chain project management* (2nd ed.). Artech House.
9. Tukel, O. I., Rom, W. O., & Eksioglu, S. D. (2006). An investigation of buffer sizing techniques in critical chain scheduling. *European Journal of Operational Research, 172*(2), 401–416.
10. Woeppel, M. J. (2006). *Projects in less time: A synopsis of critical chain*. Pinnacle Strategies Pub.
11. Rabbani, M., et al. (2007). A new heuristic for resource-constrained project scheduling in stochastic networks using critical chain concept. *European Journal of Operational Research, 176*(2), 794–808.
12. Blackstone, J. H., Cox, J. F., & Schleier, J. G. (2009). A tutorial on project management from a theory of constraints perspective. *International Journal of Production Research, 47*(24), 7029–7046.
13. Zohrehvandi, S., & Khalilzadeh, M. (2019). APRT-FMEA buffer sizing method in scheduling of a wind farm construction project. *Engineering, Construction and Architectural Management., 26*, 1129. https://doi.org/10.1108/ECAM-04-2018-0161
14. Bie, L., Cui, N., & Zhang, X. (2012). Buffer sizing approach with dependence assumption between activities in critical chain scheduling. *International Journal of Production Research, 50*(24), 7343–7356.
15. Zhang, J., Song, X., & Díaz, E. (2017). Critical chain project buffer sizing based on resource constraints. *International Journal of Production Research, 55*(3), 671–683.
16. Sarkar, D., Jha, K. N., & Patel, S. (2018). Critical chain project management for a highway construction project with a focus on theory of constraints. *International Journal of Construction Management, 21*, 1–14.

17. Zohrehvandi, S., Khalilzadeh, M., Amiri, M., & Shadrokh, S. (2020). A heuristic buffer sizing algorithm for implementing a renewable energy project. *Automation in Construction, 117*, 103267. https://doi.org/10.1016/j.autcon.2020.103267
18. Zohrehvandi, S., Vanhoucke, M., & Khalilzadeh, M. (2020). A project buffer and resource management model in energy sector; a case study in construction of a wind farm project. *International Journal of Energy Sector Management, 14*(6), 1123–1142. https://doi.org/10.1108/IJESM-10-2019-0025
19. Zohrehvandi, S., Khalilzadeh, M., Amiri, M., & Shadrokh, S. (2021). Project buffer sizing and dynamic buffer consumption algorithm in power generation construction. *Engineering, Construction and Architectural Management, 29*(2), 716–738. https://doi.org/10.1108/ECAM-08-2020-0605
20. Poshdar, M., González, V. A., Raftery, G. M., Orozco, F., Romeo, J. S., & Forcael, E. (2016). A probabilistic-based method to determine optimum size of project buffer in construction schedules. *Journal of Construction Engineering and Management, 142*(10), 04016046.
21. Bakry, I., Moselhi, O., & Zayed, T. (2016). Optimized scheduling and buffering of repetitive construction projects under uncertainty. *Engineering, Construction and Architectural Management, 23*(6), 782–800.
22. Zohrehvandi, S., & Soltani, R. (2022). Project scheduling and buffer management: A comprehensive review and future directions. *Journal of Project Management, 7*(2), 121–132.
23. Zohrehvandi, S. (2022). A project-scheduling and resource management heuristic algorithm in the construction of combined cycle power plant projects. *Computers, 11*(2), 23. https://doi.org/10.3390/computers11020023
24. Alfieri, A., Matta, A., & Pastore, E. (2016). A column generation algorithm for the buffer allocation problem approximated by the time buffer concept. *IFAC-PapersOnLine, 49*(12), 739–744.
25. Hu, X., Demeulemeester, E., Cui, N., Wang, J., & Tian, W. (2017). Improved critical chain buffer management framework considering resource costs and schedule stability. *Flexible Services and Manufacturing Journal, 29*(2), 159–183.
26. Ghoddousi, P., Ansari, R., & Makui, A. (2017). An improved robust buffer allocation method for the project scheduling problem. *Engineering Optimization, 49*(4), 718–731.
27. Zohrehvandi, S. (2022). Modeling in project planning & scheduling in construction management and project time optimization. *Academia Letters*, 4765. https://doi.org/10.20935/AL4765
28. Zohrehvandi, S., Vanhoucke, M., Khalilzadeh, M., Amiri, M., & Shadrokh, S. (2022). A fuzzy project buffer management algorithm: A case study in the construction of a renewable project. *International Journal of Construction Management*, 1–10. https://doi.org/10.1080/15623599.2022.2045860
29. Ghoddousi, P., Ansari, R., & Makui, A. (2017). A risk-oriented buffer allocation model based on critical chain project management. *KSCE Journal of Civil Engineering, 21*(5), 1536–1548.
30. Martens, A., & Vanhoucke, M. (2017). A buffer control method for top-down project control. *European Journal of Operational Research, 262*(1), 274–286.
31. Zhang, J., Jia, S., & Diaz, E. (2018). Dynamic monitoring and control of a critical chain project based on phase buffer allocation. *Journal of the Operational Research Society, 69*, 1–12.
32. Zohrehvandi, S., Soltani, R., Lefebvre, D., Zohrehvandi, M., & Tenera, A. (2022). A two-stage machine learning-based heuristic algorithm for buffer management and project scheduling optimization. *Communications in Computer and Information Science*.

Comparison of the Impacts of the Financial Crisis and the Covid-19 Pandemic Crisis on Selected Financial Indicator of Construction Companies

Katarína Krajníková, Jana Smetanková, Rastislav Ručinský, Peter Mésároš, and Lucia Zemanová

1 Introduction

The economic crisis is a phenomenon that is very difficult to stop. It is a natural course of the economic cycle, alternating between the peak of economic prosperity and the downturn. Among the most famous economic crises in the history of mankind are the Tulip Crisis (1637), the Great Depression (1929–1933), the Mexican Crisis (1994), the Asian Crisis (1997), the Russian Crisis (1998), the Brazilian Crisis (1999), Dot-com bubble (2000), Argentine crisis (2002), and Modern financial crisis (2007–2009) [1].

The accompanying phenomenon of the economic crisis is the disruption of financial markets, which will make these markets incapable of locating efficient capital where there is the greatest potential for capital appreciation [2].

The economic crisis is a sharp, immediate, ultracyclic deterioration of all or the vast majority of financial indicators – short-term interest rates, asset prices, corporate solvency, and the failures of financial institutions [3].

The financial crisis of 2008 was the biggest economic crisis since the Great Depression of 1929. The financial crisis of 2008 is called the mortgage crisis. The crisis arose as a direct consequence of the decline in liquidity in the world's financial

K. Krajníková · P. Mésároš · L. Zemanová
Technical University of Kosice, Faculty of Civil Engineering, Institute of Construction Technology, Economics and Management, Kosice, Slovakia
e-mail: katarina.krajnikova@tuke.sk; peter.mesaros@tuke.sk; lucia.zemanova@tuke.sk

J. Smetanková (✉) · R. Ručinský
Technical University of Kosice, Faculty of Civil Engineering, Expert's Institute in Construction, Kosice, Slovakia
e-mail: jana.smetankova@tuke.sk; rastislav.rucinsky@tuke.sk

L. Knapčíková, D. Peraković (eds.), *7th EAI International Conference on Management of Manufacturing Systems*, EAI/Springer Innovations in Communication and Computing, https://doi.org/10.1007/978-3-031-22719-6_15

markets, which arose in the United States as a result of the collapse of the American housing market [4].

The crisis began in March 2008, when investors sold their shares to Bear Stearns Investment Bank because it had too many so-called "toxic" assets. Bear Stearns approached JP Morgan Chase for help, but the guarantee was about $30 billion. The situation on Wall Street has been deteriorating over the summer [5, 6].

Deregulation of financial derivatives has been a key cause of the financial crisis. The laws allowed banks to invest in housing-related derivatives. These products were so lucrative that banks began to lend to riskier borrowers as well, leading to instability and a subsequent crisis. The Financial Services Modernization Act of 1999 allowed banks to use deposits to invest in derivatives [7].

There are still many disputes between economists about the root causes of this global financial crisis. In summary, the following aspects are defined as the main causes of the 2008 crisis:

- The Federal Reserve (Fed) and the United States Central Bank, which expected a mild recession beginning in 2001, reduced federal funds rates (interest rates that banks charge each other for federal funds loans). This decline has allowed banks to extend consumer credit at a lower primary rate and has encouraged them to lend even "subprime" or high-risk loans, albeit at higher interest rates. Consumers have used cheap loans to buy to-wares – appliances, cars, and houses. The result was the so-called bubbles that have significantly exceeded the basic or intrinsic value.
- Changes in banking laws, have made it possible to offer clients subprime mortgages and mortgage loans that have been structured into unusually high payments with a maturity at or near the end of the loan. or with adjustable interest rates, where the rate was fixed at relatively low levels during the initial period and subsequently fluctuated at the level of federal funds rates. The result of this process was that the share of the so-called subprime mortgages increased from around 2.5% to almost 15% per year among all housing loans from the 1990s to 2004–2007.
- Widespread securitization, in which banks combined hundreds or even thousands of high-risk mortgages with less risky forms of consumer debt and sold them on the capital markets as securities (bonds) to other banks and investors, including hedge funds, also contributed to the growth of subprime mortgages and pension funds.
- In 1999, the Depression-era Glass-Steagall Act (1933) was partially repealed, which allowed banks, insurance companies, and securities companies to enter and merge with each other, resulting in banks that were too so great that they could jeopardize the entire financial system. At the same time, in 2004 the Securities and Exchange Commission (SEC) weakened the value of net capital (the ratio of capital or assets to debt or liabilities that banks are required to maintain as protection against insolvency).
- The long period of global economic stability and growth that immediately preceded the crisis has convinced many banking executives, officials, and

economists that extreme economic volatility is a thing of the past. This approach, together with the ideology emphasizing the regulation and ability of financial firms to self-control, has led to a loss of attention and the consequent neglect of the first signs of the impending crisis [4, 8–11].

At the end of 2019, the World Health Organization (WHO) was informed of cases of pneumonia of unknown cause in Wuhan, China. Subsequently, the Chinese authorities (January 7, 2020) identified a new coronavirus called Covid-19 as the cause of the inflammation. The new coronavirus is a new strain that has not yet been identified in humans. On January 30, 2020, the WHO declared a Public Health Emergency of International Importance (PHEIC), the highest level of WHO pop alarms. In 18 countries outside China, 98 cases and no deaths were reported. On March 11, 2020, following an enormous increase in cases outside China, the WHO declared that the outbreak could be described as a pandemic. To date, more than 118,000 cases have been reported in 114 countries and 4291 deaths have been reported. In mid-March, Europe became the epicenter of the pandemic, reporting more than 40% of confirmed cases worldwide [12].

The Covid-19 pandemic is much more than just a health crisis, affecting societies and economies at their core. The Covid-19 pandemic has led to dramatic loss of life around the world and poses an unprecedented challenge to public health, food systems, and the world of work. The social and economic disturbances caused by the pandemic have affected a wide range of societies. Tens of millions of people were in danger of falling into extreme poverty. Assessing the impact of the Covid-19 crisis on societies, economies and vulnerable groups is essential to informing and adapting the responses of governments and partners to recovering from the crisis and ensuring that no one is left behind in this effort. Without socio-economic responses, the risk to life and livelihoods will increase in the coming years. Immediate development responses to this crisis needed to be pursued with a view to the future [12, 13].

2 Literature Review

The financial crisis in 2008 affected all businesses. Many small businesses have failed or been forced to lay off employees, cut spending, halt expansion plans, and find new ways to survive. The number of companies created annually in the decade before the financial crisis averaged 670,000 per year, reaching a maximum of more than 715,000 in 2006. Initial numbers fell dramatically during the crisis, reaching a minimum in 2010 (560,000 companies). Between December 2008 and December 2010, about 1.8 million small businesses disappeared [14].

Small businesses are referred to as "job creators", but during the financial crisis, redundancies in small businesses were dramatic. In 2 years (December 2007–December 2009), about 8.7 million jobs were lost. During the recession, small businesses did not create jobs, about 60% of employees lost their jobs. Ten years after the crisis, small businesses have returned to creating around 60% of all new jobs. The financial crisis hit the financial sector in particular, where before the financial crisis the number of commercial loans of small businesses was relatively high, the traditional possibility of lending was still growing at almost double digits. This stopped after the outbreak of the financial crisis. In the years 2008 to 2011, large banks' loans to small businesses were virtually non-existent. The total amount of commercial loans to small businesses fell by $ 40 billion between the second quarter of 2008 and the second quarter of 2010.

The Effects of the Crisis on the Financial Performance of Malaysian Companies (Ben Chin Fook Yap, Zulkifflee Mohamed, K-Rine Chong) examined the impact of the crisis on the financial performance of 70 companies in the manufacturing sector over a period of 5 years, specifically from 2006 to 2010. Using factor analysis, the initial set of 21 financial indicators was reduced to 6 significant indicators: liabilities, profit before tax, total debts, cash flow, total asset turnover, and inventory. Based on the above indicators, the companies were divided into 4 categories, namely weak, below average, above average, and companies with good financial results. The results showed that the financial crisis had a direct impact on the finances of the companies included in the study. Of the selected companies, 46 companies qualified as well-financed in 2006, while in 2010 only 6 companies were defined in this way. It was interesting to note that up to 27 companies fell into the non-profit/lossy companies' category during the period, which was an increase of 20 companies compared to year's 2006. As many as 15 of the 17 companies in the average categories fell into the poor performance category. A key finding of this study was that in the event of a financial or economic crisis, the finances of most companies are seriously or adversely affected, and if negative economic conditions do not improve, there is a high probability that many companies will face liquidity and solvency problems, which could ultimately lead to collapse and bankruptcy [14, 15].

The beginning of 2020 was for many companies one of the most challenging in their existence. Following the outbreak of the global financial crisis, the market has been recovering slowly and economic growth has been recorded in all industries in recent years. Therefore, even in the beginning of 2020, continued economic growth was recorded. However, everything changed in mid-March due to the outbreak of the Covid-19 pandemic in most parts of the world. Many governments have responded to restrictions on economic activities, such as the cessation of production, the blocking of certain sectors and activities. The research "how COVID-19 infects financial reporting and results presentation – Q1 2020" addressed companies from the United Kingdom, Switzerland, and Germany operating in the pharmaceutical industry, chemical industry, consumer software production, transport and logistics, telecommunications, tourism, and other services.

Although the sample of companies in the above statistical study was not representative, it showed a clear impact of the Covid-19 pandemic on individual sectors. Consumer software production (changes in revenues – decrease by 6%, EBITDA – decrease by 52%), tourism (changes in revenues – decrease by 13%, EBITDA – decrease by 496%) were blocked and affected immediately. The mining/commodity industry (changes in revenues – a decrease of 10%, EBITDA – a decrease of

166%) was mainly influenced by a sharp drop in oil prices and significantly weaker demand. The automotive industry (revenue changes – down 5%, EBITDA – down 54%) suffered from a sharp drop in demand, especially in the important Asia-Pacific region. On the contrary, it was recorded at once in the field of science (change in income – increase by 8%, EBITDA – increase by 30%) and the telecommunications sector (EBITDA – increase by 2%). The study also predicted potential effects, either short-term, such as closing deals, or long-term, such as changing business models [16].

Both of these crises have had profound adverse effects at the global level. The impact of the global financial crisis and the Covid-19 pandemic was examined in A Comparative Analysis of COVID-19 and Global Financial Crises: Evidence from the US Economy (Zongyun Li, Panteha Farmanesh, Dervis Kirikkaleli, Ranita Itani), from the beginning of 2021, on the macroeconomic variables of the US economy. The research points out that the current crisis caused by the Covid-19 pandemic cannot be examined empirically and quantitatively, given its ongoing nature. The authors also point to the fact that no detailed study comparing the global financial crisis and the crisis caused by the Covid-19 pandemic is available. The study thoroughly explains the financial crisis and the impact of Covid-19 on the macroeconomic variables of the US economy. The study analyzed each variable, calculating the average change and the average of each variable. Industrial production, consumer spending, overall trade, and unemployment are severely affected in the current pandemic crisis. The survey showed a greater impact of the pandemic on selected aspects than during the global financial crisis. The probability of a smooth recession was high during the global financial crisis due to the financial crisis and the financial collapses of individual banks and investment companies. Research has shown that it is not uncommon for the economy not to fall into recession during a crisis. Such a crisis is adversely affecting stock levels and affecting international trade, leading to a decline in industrial production and employment levels. Unemployment then further creates room for low consumer spending [11].

The global financial crisis in 2008 had a relatively strong impact on production in industry and construction. Between February and April 2008, the index for total construction in the EU fell by 7.2 points. In the following months of 2008, the ideox fell again, although the declines were generally smaller and there were occasional increases. However, over the next 5 years (until spring 2013), the construction index had a clearly declining trend. During these years, the overall construction index lost 33 points (the lowest level of 95 points – March 2013). Subsequently, the sector slowly recovered and returned to its previous highs. However, from February to April 2020, the construction idnex fell by 29 points in the EU and by 34 points in the euro area. In other words, during the first wave of the Covid-19 crisis, the construction index lost about as much as during the 5 years of the economic and financial crisis [17].

In the Slovak construction industry, after years of growth, there was a slight decrease in construction output, specifically to the level of 5253 mil. EUR, which represents a year-on-year decrease of about 9.9%. Domestic construction output was focused mainly on new construction, reconstruction, and modernization, where an increase of 3.2% was recorded compared to the previous year. Construction production in the area of repairs and maintenance in the Czech Republic also increased, namely by 1.5% compared to the previous year. Data on construction production abroad are interesting. Despite many limitations, the increase in construction output is also perceived in the mentioned section, specifically by 2.5% compared to the previous year (Statistical Office of the Slovak Republic, 2021). The capacities of construction companies are currently occupied at 75%. In the second half of this year, capacity utilization should increase slightly, to 84% on average [18–20].

A number of studies have been carried out around the world focusing on the effects of the great financial crisis and the effects of the Covid-19 pandemic on individual aspects of industry and the activities of companies. These surveys point to the significant effects of these crises on the activities of organizations and their economic results. Based on the above, the aim of the research is to analyze and compare the effects of global crises, the global financial crisis, and the crisis caused by the Covid-19 pandemic, on the profitability of sales and the economic results of selected construction companies.

3 Research Methodology and Data

3.1 Research Aim

The research methodology consisted of analysis and mutual comparison of profitability of sales and economic results of selected construction companies over three periods, namely:

- The period of the financial crisis in 2008.
- Transitional period – market stabilization after the financial crisis – year 2014.
- The period of crisis caused by the Covid-19 pandemic – 2020.

The aim of the research was to find out whether the impact of the Covid-19 pandemic on the profitability of companies' revenues is greater than the impact of the economic crisis in 2008.

Economic parameters were obtained from the final annual reports of the companies and from the publicly available economic database FinStat. Data were analyzed using the Kolmogorov directionality test, the Friedman repeated measures test, the Wilcoxon Signed-Rank test, and the Mann-Whitney U test.

3.2 Sample and Data Selection

The main obstacle in the research was the availability and quality of data. The data used in the research were obtained from the annual reports of selected companies. Annual reports on the company's activities were obtained from the publicly available web database of Finstat.sk companies. The FinStat database is a data source that provides information on selected financial indicators of selected companies. The respondents selected for this study were selected on the basis of the Eurostav TOP 100 ranking, which is published annually in Slovakia. The analysis compared 30 companies. These companies are among the companies with a long tradition and a stable market share. As part of the analysis, companies focused on the exhibition of building and civil engineering were analyzed, namely 13 companies specializing in the construction of buildings and 17 companies specializing in the construction of civil engineering.

3.3 Methodology

To achieve the above research goal, the following steps were taken:

- A review of the literature on the global financial crisis and the crisis caused by the Covid-19 pandemic: defining the causes, impacts, and an overview of the research carried out.
- Impacts of selected crises on the construction industry – summary of surveys carried out.
- An overview of the current state of construction in the European Union and in the territory of the Slovak Republic.
- Determination of basic hypotheses.
- Data analysis, definition of conclusions, and future recommendations.

Based on the research goal, the following hypotheses were established:

- *Hypothesis 1*: Profitability of sales is the same in all years (2008, 2014, and 2020).
- *Hypothesis 2*: The crisis caused by the Covid-19 pandemic and the financial crisis in 2008 had the same impact on income profitability.

4 Results

The main aim of the research was to analyze whether the impact of the Covid-19 pandemic (2020) on the profitability of companies is greater than the impact of the economic crisis (2008). Data were collected from 30 companies for which the economic result and sales for own services and goods were defined in individual years, i.e., the period 2008, 2014, and 2020. The data were obtained from the FinStat database and the final annual reports of selected companies. By profitability we mean the share of the economic result from economic activity and revenues for own services and goods.

Table 1 Results of Kolmogorov-Smirnov test of normality

Year	2020	2014	2008
Result	The value of the K-S test statistic (*D*) is 0.31746	The value of the K-S test statistic (*D*) is 0.34213.	The value of the K-S test statistic (*D*) is 0.28524.
	The p-value is <0.00001.	*The p-value is <0.00001*	*The p-value is 0.00008.*
	Rejected hypothesis H0	*Rejected hypothesis H0*	*Rejected hypothesis H0*
	Provides very good evidence that, data *not* normally distributed	Provides very good evidence that, data *not* normally distributed	Provides very good evidence that, data *not* normally distributed
Result details			
Mean	32.97	237.33	497.92
Median	3.5	8.5	28
Standard deviation	70.02	480.01	885.05
Skewness	3.73	2.69	2.63
Kurtosis	17.29	7.59	7.85

As a first step, we examined the normality of the data obtained. We used the Kolmogorov-Smirnov test of normality and partial hypotheses:

- *H0*: data is from normal distribution.
- *H1*: data not from normal distribution.

We found that in neither case could we consider the data distribution to be normal. Summary results are displayed in Table 1.

In the next step, the Friedman test for repeated measurements was used, which is a suitable non-parametric test for comparing the statistical significance of averages in individual years. It is used to determine whether or not there is a statistically significant difference between the averages of three or more groups in which the same subjects occur in each group. The following sub-hypotheses were established:

- *H0*: the averages of profitability of sales are the same in all years.
- *H1*: the averages of profitability of sales are not the same in all years.

The results of this statistical test were as follows:

The X^2 statistic is 6.7583. The result *is significant* at $p < 0.05$.

The p-value is 0.03408, therefore, I reject H0, which means that the average in at least 1 year is statistically significantly different from other years.

The Wilcoxon Signed-Rank test, which is a non-parametric variant for comparing the averages of 2 dependent samples, was chosen as the post-test. It is based on difference scores, but in addition to analyzing the signs of the differences, it also takes into account the magnitude of the observed differences.

- *H0*: average profitability of sales is the same in both years.
- *H1*: average profitability of sales is not the same in both years.

Table 2 *Výsledky* results of Friedman's test for repeated measures

Comparable periods	2020 vs. 2014	2014 vs. 2008	2020 vs.2008
Result	The value of *z* is −1.362.	The value of *z* is −3.3012	The result is significant at $p < 0.05$.
	The p-value is 0.17384	*The p-value is 0.00096.*	*The p-value is 0.02926*
	Do not rejected H0	*Rejected hypothesis H0*	*Rejected hypothesis H0*
	The averages in these years *are not statistically significantly different.*	The averages in these years *are statistically significantly different.*	The averages in these years *are statistically significantly different.*
	The result *is not* significant *at p < 0.05.*	The result *is* significant *at p < 0.05.*	The result *is* significant *at p < 0.05.*
Result details			
W-value	178.5	72	137
Mean difference	570.03	−1337.33	−1115.77
Sum of pos. ranks	317.5	72	137
Sum of neg. ranks	178.5	393	359
Z-value	−1.36	−3.30	−2.18
Mean	248	232.5	248
Standard deviation	51.03	48.62	51.03

Table 3 Results – the impact of crises on corporate profitability (Mann-Whitney U test)

Comparable periods	2020 vs. 2008	2014 vs. 2008
Result details		
Sum of ranks	3652	3608
Mean of ranks	*60.87*	*60.13*
Expected sum of ranks	3630	3630
Expected sum of ranks	60.5	60.5
U-value	1778	1822
Expected *U*-value	1800	1800

The results are shown in detail in Table 2.

To clearly determine from the examined data which of the years (2020 or 2014) had a greater impact on companies' profitability, the non-parametric Mann-Whitney U test, which is a suitable test for comparing two independent sets of measurements was used. The following hypotheses were established:

- *H0*: the averages of the differences in the profitability of sales are the same in both years.
- *H1*: the averages of the differences in the profitability of sales are not the same in both years.

Result details are in Table 3.

The Z-Score is 0.11285. The *p-value is 0.9124*. The result *is not significant* at $p < 0.05$, *H0* was not rejected, which means that for our selection of companies, the crisis caused by the Covid-19 pandemic and the financial crisis in 2008 had *the same* impact on their profitability.

5 Discussion and Recommendations

We live in an unstable time today. The company and the financial market gradually began to recover from the effects of the Covid-19 pandemic on individual segments of society, and there was a new fact, a war conflict in Ukraine, which is gradually crippling the entire market and society. Even before the outbreak of the war and after years of very low inflation, inflation reached its highest level in 13 years in the second half of 2021. The main causes of inflation are the re-opening of the economy, inflationary pressures due to higher energy prices, and the so-called basic effects [21].

In view of these facts, another economic crisis is increasingly escalating. In order for companies to eliminate the effects of possible crises as much as possible, it is necessary to take certain steps and follow them even if the crisis occurs. These are mainly the following steps:

- The need to increase the creation of financial reserves – companies need to think more critically about their spending.
- Prompt repayment of their liabilities – in the event that companies report such liabilities, it is necessary that they repay them as soon as possible and not borrow further loans and borrowings.
- Employment insurance – in the event of a reduction in the number of employees as a result of the crisis, it is necessary to take out such insurance.
- Insurance of installments of your liabilities – in the case of a mortgage or other loan, it is necessary to consider their insurance, thus eliminating the risk of liquidity loss.
- Creation of a strategy, preparation of a crisis plan – it is necessary to develop a crisis plan in the event of a reduction in income or loss of employment. The company will establish a so-called "worst case scenario" and will choose steps and methodologies to eliminate and minimize possible crisis situations.

6 Conclusion

Economic crises are fluctuations in economic activity that recur at regular intervals. In history, mankind has encountered a number of global crises, such as the Tulip Crisis in 1637 or the Great Depression in the pre-war period. These and many other crises have greatly affected the global economy and the lives of millions of people. In the twenty-first century, the company faced two major crises, the financial crisis, which began in 2007 and lasted for 2 years, and the crisis caused by the Covid-19 pandemic. These crises have severely crippled all aspects of society. The paper analyzed how individual crises affected the activities of selected construction companies operating in the Slovak Republic. The research analyzed the results of management, the respective profitability of sales of selected construction companies, which are among the leaders in this segment of the economy. The survey showed that both global crises (the financial crisis and the Covid-19 pandemic) had an impact on the development of construction companies' revenues and confirmed the hypothesis that there was a demonstrable link within selected companies, both crises had the same impact on revenue profitability. The effects of the crises reflect tensions and imbalances, i.e. partial nerve balances between the previous period and the present. Within this fact, the goal of each company is to prepare a consistent strategic crisis plan and create reserves that can help eliminate adverse impacts and help prompt and effective implementation of changes.

Acknowledgments The paper presents partial research results of project APVV-17-0549 "Research of knowledge-based and virtual technologies for intelligent designing and realization of building projects with emphasis on economic efficiency and sustainability".

References

1. Fekete, D. (2021). *TOP 9 biggest crises in human history*. Retrieved from https://financer.com/sk/blog/top-9-najvacsich-kriz-v-historii-ludstva/. Accessed 17 June 2022.
2. Lisý, J. (2005). *Economic performance and economic growth* (2nd ed., p. 132). IURA Edition. ISBN 80-8078-035-8.
3. Euroekonom.sk. (2021). *The economic crisis and the cyclical development of the economy*. Retrieved from https://www.euroekonom.sk/ekonomika/hospodarska-kriza/. Accessed 17 June 2022.
4. Duignan, B. *Financial crisis of 2007–2008*. Retrieved from https://www.britannica.com/event/financial-crisis-of-2007-2008. Accessed 17 June 2022.
5. Department of the Treasury. (2021). *The financial crisis response in charts*. Retrieved from https://www.treasury.gov/resource-center/data-chart-center/Documents/20120413_FinancialCrisisResponse.pdf. Accessed 10 June 2022.
6. Brunnermeier, M. K. (2008). Deciphering the liguqidity and credit crunch 2007–2008. *Journal of Economic Perspectives, 23*(1), 77–100. Retrieved from https://pubs.aeaweb.org/doi/pdfplus/10.1257/jep.23.1.77. Accessed June 2022
7. U.S. Government Publishing Office. (2020). *Grammn Leach Bliley Act-Public Law 106–102-NOV.12 1999*. Retrieved from https://www.govinfo.gov/content/pkg/PLAW-106publ102/pdf/PLAW-106publ102.pdf. Accessed 30 November 2021.
8. Singh, M. (2022). *The 2007–2008 financial crisis in review*. Retrieved from https://www.investopedia.com/articles/economics/09/financial-crisis-review.asp. Accessed 15 June 2022.
9. Amadeo, K. (2022). *2008 financial crisis*. Retrieved from https://www.thebalance.com/2008-financial-crisis-3305679. Accessed 15 June 2022.
10. Britanica. (2022). *Key events of the crisis*. Retrieved from https://www.britannica.com/event/financial-crisis-of-2007-2008/Key-events-of-the-crisis. Accessed 12 June 2022.
11. Li, Z., Farmanesh, P., Kirikkaleli, D., & Itani R. A comparative analysis of COVID-19 and global financial crises:evidence from US economy. *Economic Research – Ekonomska Istraživanja*. Received 1 April 2021, Accepted 3 July 2021, https://doi.org/10.1080/1331677X.2021.1952640. Retrieved from https://www.tandfonline.com/doi/full/10.1080/1331677X.2021.1952640. Accessed 11 June 2022.
12. WHO. (2022). *Coronavirus Covid-19*. Retrieved from https://www.euro.who.int/en/health-topics/health-emergencies/coronavirus-covid-19/novel-coronavirus-2019-ncov. Accessed 14 June 2022.

13. UNDP.org. (2021). *Covid-19: Socio-economic impact*. Retrieved from https://www.undp.org/coronavirus/socio-economic-impact-covid-19. Accessed 8 June 2022.
14. Weltman, B. (2022). *10 years after the financial crisis: The impact of small business*. Retrieved from https://www.investopedia.com/small-business/10-years-after-financial-crisis-impact-small-business/. Accessed 5 June 2022.
15. Chin Fook Yap, B., Zulkifflee, M., & Chong, K.-R. (2014). The effects of the crisis on the financial performance of Malaysian companies, Macrothink Institute. *Asian Journal of Finance & Accounting, 6*(1), 236–248. ISSN 1946-052X. Retrieved from https://www.researchgate.net/publication/262689062_The_Effects_of_the_Crisis_on_the_Financial_Performance_of_Malaysian_Companies. Accessed 30 May 2022
16. Köster O., & Igoe, S. (2020). *Perspectives: How COVID-19 infects financial reporting and results presentation – Q1 2020*. Deloitte. Retrieved from https://www2.deloitte.com/ch/en/pages/audit/articles/financial-reporting-survey-q1-2020.html. Accessed 18 May 2022.
17. Eurostat. *Impat of Covid-19 crisis on construction*. Retrieved from https://ec.europa.eu/eurostat/statistics-explained/index.php?title=Impact_of_Covid-19_crisis_on_construction#Comparison_with_2008. Accessed 18 May 2022.
18. CEEC Research. (2021). *Half-year study of Slovak construction H1/2021*. Retrieved from https://www.archinfo.sk/diskusia/blog/vykon-povolania/polrocna-studia-slovenskeho-stavebnictva-h1-2021-ceec-research.html. Accessed 4 June 2022.
19. SR Statistical Office of the Slovak Republic. (2021). *Industry statistics: Construction-construction production*. DATAcube. Retrieved from http://datacube.statistics.sk/#!/view/sk/VBD_INTERN/st0001ms/v_st0001ms_00_00_00_sk. Accessed 4 June 2022.
20. Smetanková, J., Mésároš, P., & Ďuriš, A. (2021). Results of Slovak construction companies in 2020. *Eurostav: Professional-Scientific Peer-Reviewed Journal on Construction and Architecture, 27*(7–8), 91. 2021 /1335-1249/.
21. ECB. (2021). *Why is inflation so high at the moment?* Retrieved from https://www.ecb.europa.eu/ecb/educational/explainers/tell-me-more/html/high_inflation.sk.html. Accessed 18 June 2022.

Part IV
Industry 4.0

Opportunities of Using Machine Learning Methods in Telecommunications and Industry 4.0 – A Survey

Dragan Peraković, Marko Periša, Ivan Cvitić, Petra Zorić, Tibor Mijo Kuljanić, and David Aleksić

1 Introduction

Machine learning (ML) is a subsystem of artificial intelligence (AI) used in computer science. It is focused on building programs and applications that will learn from available data and improve their accuracy over time without being programmed to do so. One of the main characteristics of ML is learning from experiences or data sets, not just from instructions.

The research presented in this paper aims to analyze the possibility of applying ML in telecommunications and Industry 4.0. A prerequisite for achieving the stated goal of this work is knowledge of the definition of ML and the principles of work, as well as the advantages and disadvantages of its application.

The paper outline follows, first of all, the term AI and its application will be defined. The above is a starting point for understanding ML's concept, work, and impact. Intelligence and AI are divided into subsystems that are counted under AI. Also, some of these subsystems will be briefly described. Furthermore, the paper will explain the working principle of ML, i.e., how machines learn and all other steps that must be completed before the machine is ready for use. ML is divided into supervised, unsupervised, and reinforced ML. The division of ML into the listed types will be described according to their working principle and features, as well as the differences between them.

D. Peraković (✉) · M. Periša · I. Cvitić · P. Zorić · D. Aleksić
University of Zagreb, Faculty of Transport and Traffic Sciences, Zagreb, Croatia
e-mail: dragan.perakovic@fpz.unizg.hr; marko.perisa@fpz.unizg.hr; ivan.cvitic@fpz.unizg.hr; petra.zoric@fpz.unizg.hr

T. M. Kuljanić
Hrvatska Elektroprivreda d.d, Zagreb, Croatia

L. Knapčíková, D. Peraković (eds.), *7th EAI International Conference on Management of Manufacturing Systems*, EAI/Springer Innovations in Communication and Computing, https://doi.org/10.1007/978-3-031-22719-6_16

Also, the paper will show the division into tools and platforms used during the ML process. A comparison was made between the tools according to their essential characteristics, and two examples were presented for the K-means clustering process in two of the available platforms.

ML and AI in the field of telecommunications can be applied to an extensive range of services and currently available systems. One of ML's roles is protecting and securing the network from cyber-attacks. In addition to its role in network security, the paper also defines the method of applying ML in classifying Internet of Things (IoT) devices. Some other examples of the application of ML are listed along with a graphic presentation of the workflow of ML in the field of Internet and telecommunications.

2 Overview and Analysis of Machine Learning and Artificial Intelligence Concepts

AI refers to areas of computer science that involve observing and developing intelligent systems, tools, and machines that react and learn like humans. AI systems are designed to perceive the environment through data collection, interpret the collected data, which does not necessarily have to be structured, make judgments based on previous knowledge or processing of information obtained from the data, and evaluate the best actions to be taken to achieve the given goal [1].

ML is a branch of AI that deals with designing algorithms that improve their efficiency based on empirical data. It can be divided into three main types: supervised, unsupervised, and so-called reinforcement learning.

2.1 Supervised Machine Learning

Supervised machine learning is a form of ML that is used quite often. Supervised learning works on the principle that we get an input variable or data in general, and the task is to predict the exact output data or label. ML supervisory algorithms are designed to work on the principle of learning by example. When training this type of algorithm, the person managing the training (supervisor) must provide specific input data with the correct output data. During training, the algorithm looks for specific repeating patterns so that it can later know the output itself without the help of a supervisor. Supervised learning can be divided into two subtypes: classification and regression [2].

In the classification type of supervised machine learning, classification algorithms obtain data from a specific group. Their task is to classify these data into specific groups or classes into which they would fall based on the training. One of the most famous examples of this type of supervised machine learning is used in electronic mail when determining incoming messages. These messages are classified into specific categories, such as spam messages. Regression is a predictable statistical process where a model tries to find important links between dependent and independent variables. These algorithms are used to predict sales numbers, revenue generated, exam or test results, etc. The three most common regression algorithms are linear, logistic, and polynomial [2].

2.2 *Unsupervised Machine Learning*

With unsupervised machine learning, output labels or exact output data are not known. The task of this ML method is to discover the structure of the output data. An unsupervised machine learning model cannot be trained. That is, we cannot teach it what the output variable will be when even the team working on ML of this type does not know what it could be. For this reason, machines, or models, have to look for specific hidden patterns and information that are most likely not even visible to the human eye to learn something. This is a process where models learn on their own. Unsupervised machine learning has its type of algorithm called clustering. Clustering algorithms can be divided into exclusive, overlapping, hierarchical, and probabilistic [3].

Exclusive clustering works on the principle that if specific data belongs to one group or cluster, it cannot be included as a member of any other cluster. The opposite is overlapping clustering, where data can belong to two or more clusters. Hierarchical clustering is based on the union between two neighboring clusters, and probabilistic clustering is based on pure mathematical laws of probability. These four clustering algorithms have their subtypes, of which the most commonly used are K-means and hierarchical clustering [3].

The differences in the use of supervised and unsupervised machine learning are that supervised learning is used for situations where we know exactly what we are looking for in the data, and we know the results from which we can teach the models how to work and how to approach problems, while with unsupervised machine learning we do not know what we are looking for and what the result or output could be. Because of this, we cannot even teach the models, but we have to let them come to conclusions based on their algorithms that will help them learn particular things independently. They work based on their own experience gained during the research process. For these reasons, unsupervised learning has the problem of possibly giving results with a higher possibility of error [4].

Unsupervised machine learning can be used in data preparation for supervised machine learning, where the task is to detect output labels and data. It can also be used in recommendation systems that try to predict specific data, such as the rating a user has given to a product. Unsupervised learning can still be used to detect anomalies, i.e., to detect some rare objects or events that have certain deviations in some respect from the group in which they are located [5].

2.3 Reinforcement Machine Learning

Reinforcement machine learning is a type of machine learning that deals with the research of how intelligent agents should behave, how they should perform actions in a specific environment, and how to reach the best possible outcomes with these actions [6].

As with previous types of ML, reinforcement machine learning has its algorithms: value-based, policy-based, and model-based. Reinforced machine learning can be divided into positive reinforcement machine learning and negative reinforcement machine learning. Positive reinforced machine learning is a type of reinforcement learning in which the strength and frequency of behavior increase due to specific behavior. Its advantages are that maximized performance and changes remain for a longer time. The downside is that the results can be diluted if there is too much magnification. Negative reinforcement machine learning is the reinforcement of behavior that occurs due to a negative state meant to be stopped. The downside of this type of reinforcement machine learning is that it can only achieve minimal behavior [7].

This form of ML has a wide application in the world, so machine learning strengthened in this way can be found in robotics, traffic light control (for now only in simulations), the gaming industry, etc.

3 Tools and Platforms for Applying Machine Learning Algorithms

Choosing the right tool for applying ML can be just as important as choosing the suitable algorithm for ML [8]. Differences in specific tools can be in features such as the platform's price, the type of language used to perform ML, the types of algorithms and features supported, and the platform on which the tool runs. Thus, some of the most famous tools for applying ML algorithms are currently: Scikit Learn, PyTorch, TensorFlow, Knime, etc. [9].

3.1 Scikit Learn Programming Platform

Scikit Learn is one of the better-known ML tools. It is a tool that provides ML development in the Python programming language. It is a simple, free, and effective tool suitable for use in predictable statistical analyses of data. Algorithms and features that make Scikit Learn work include classification, regression, clustering, dimensionality reduction, data preprocessing, and model selection. Scikit Learn supports supervised and unsupervised machine learning [9, 10]. The programming

```
import matplotlib.pyplot as plt
import numpy as np
import tensorflow as tf

points_n = 200
clusters_n = 3
iteration_n = 100
```

Fig. 1 Importing libraries in TensorFlow [12]

```
points = tf.constant(np.random.uniform(0, 10, (points_n, 2)))
centroids = tf.Variable(tf.slice(tf.random_shuffle(points), [0, 0], [clusters_n, -1]))
```

Fig. 2 Random data generation and centroid selection [12]

languages supported on this platform are Python, C++, C, and the Cython programming language.

3.2 *TensorFlow Machine Learning Platform*

TensorFlow is one of the most famous platforms for ML. According to the comparison with PyTorch, it is almost the same regarding the platforms on which it can work, the programming languages it supports, and the fact that it is also a free open-source platform. This platform is convenient. After all, it offers a simple way to create models for the training phase in ML because it has a large number of options when creating models, which allows the user to create exactly the model he needs. Also, TensorFlow offers a JavaScript library for data flow programming [9, 11].

Figure 1 shows the K-means clustering procedure using the Tensor-Flow library. The *matplotlib* and *numpy* libraries are imported, and the *pandas* library is not used. It also shows the number of groups, 3, the amount of data, 200, and the number of repetitions, 100 [12].

After importing the required libraries, the next step is to build the clustering model. First, random data is generated from which points for initial centroids are randomly taken. These lines of code are shown in Fig. 2.

The final stage of K-means clustering in TensorFlow starts with creating a graphic display. Figure 3 shows the phase of graph creation, in which the new centroids from the previous step are first updated, and their values are returned together with the cluster values.

In the end, a part of the code related to the graphic printout of the results is written. The coordinates of the centroid of each data group are displayed here.

```
with tf.Session() as sess:
  sess.run(init)
  for step in xrange(iteration_n):
    [_, centroid_values, points_values, assignment_values] = sess.run([update_centroids,
centroids, points, assignments])
```

Fig. 3 Return values to data groups and update centroids [12]

```
print("centroids", centroid_values)

plt.scatter(points_values[:, 0], points_values[:, 1], c=assignment_values, s=50, alpha=0.5)
plt.plot(centroid_values[:, 0], centroid_values[:, 1], 'kx', markersize=15)
plt.show()
```

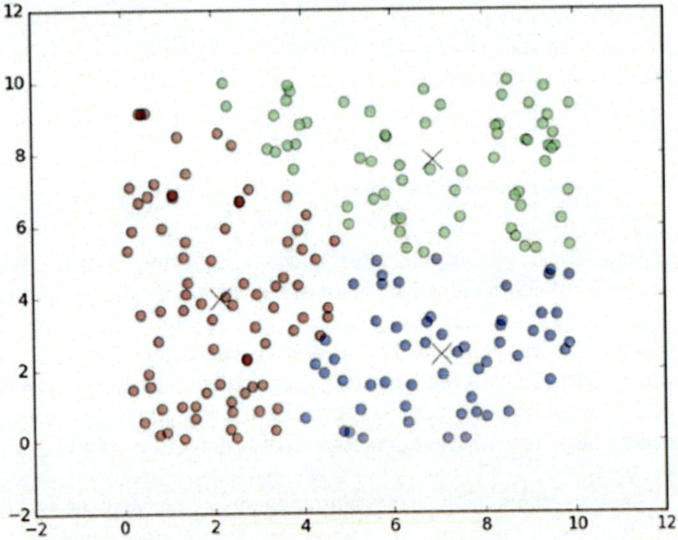

Fig. 4 The code for displaying the graphical solution and the graphical solution

Figure 4 shows the code for creating and displaying the graph, and below the code is the printout, the required graphic display.

3.3 PyTorch Framework for Machine Learning

PyTorch is also a free tool for applying ML algorithms. It differs from Scikit Learn in that the programming languages used in this tool are: Phyton, C++, and a computer platform and programming interface model called CUDA. PyTorch, like Scikit, runs on Linux, MacOS, and Windows operating systems. PyTorch is also available and has support for working on cloud services. PyTorch is a more suitable program for deep learning than ML, but it is also possible to do an ML process on this platform [9, 13].

3.4 Weka Programming Tool

Weka is also open-source software that provides tools for data preparation and the use of several ML algorithms. It also contains tools for working with regression, clustering, visualization, and classification algorithms. Weka is a free platform that runs on Windows, Linux, and MacOS operating systems. It is mostly used for data mining. It is a form of data analytics where large amounts of data are extracted for meaningful insights. Various machine learning algorithms are used for data mining, which can be used directly on the data or via Java code in the Weka program. Unlike other tools, Weka supports only the JavaScript programming language [14, 15]. What distinguishes Weka from other ML tools is that it is a tool that uses a graphical user interface (GUI) that allows loading a dataset, running, and working with algorithms [16].

3.5 RapidMiner Machine Learning Platform

This tool enables ML, deep learning, data mining, predictive analytics, etc. It can be used for research, application development, and education. It uses Java exclusively as a programming language. It can work on various operating systems. It is used to prepare data, visualize, and validate and optimize models. The difference between it and the other previously mentioned tools is that, in addition to its free version, it also offers paid versions, i.e., extensions of the tool [9]. RapidMiner, like Weka, uses a GUI that helps with implementation and design in the analytical parts of the work. RapidMiner offers the possibility of working in its extension on the Internet browser, which is increasingly popular today and enables the operation of the RapidMiner Go program. However, there is also the option of downloading the complete RapidMiner Studio program, which offers more services than the RapidMiner Go version [17].

4 Application of Machine Learning in Telecommunications

ML and AI in the field of telecommunications have a broad scope of application. Among other things, they are used to protect computer networks from unauthorized access to the network and classify IoT devices. Also, ML can be used in managing network traffic and predicting analysis results [18].

4.1 The Role of Machine Learning in Network Traffic Security

Increasingly, network traffic is under the attack of illegal attacks and attempts to steal data. Thus we have distributed denial of service (DDoS) attacks as one of the most common threats to network traffic. DDoS is a type of attack on a specific website. The goal of a DDoS attack is to flood that site with more traffic than the server or network can handle. This is done so that the traffic sent to the network consists of constant incoming messages, connection requests, fake packets, etc. With DDoS attacks, hackers achieve that the site or server they attacked becomes useless or unavailable [19].

Incoming traffic is divided into two groups: legitimate and illegitimate traffic. The types mentioned above of data sent to flood the network belong to illegitimate or fake traffic. This type of traffic includes subtypes of DDoS traffic. DDoS traffic differs by the type of protocol that was used to carry out the attack, so we have DNS DDoS attacks that use the DNS protocol, UDP DDoS attacks, CharGen DDoS attacks, and others [20].

The UDP protocol is often used in DDoS attacks due to its simplicity. The device that receives UDP packets does not have a large enough capacity to receive such a large amount of incoming traffic, and it also tries to respond with a large number of Internet Control Message Protocol (ICMP) "*destination host unreachable*" messages that create additional network congestion. The DNS protocol has been one of the leading protocols for amplifying DDoS attacks in recent years. In this type of DDoS attack, amplifiers are also used in addition to standard equipment, which are devices located outside the botnet network that provide answers to queries. Computers within the botnet network send queries and spoof the IP address of the source, i.e., the IP address of the server being attacked, and as a result, a response is sent from the booster to the IP address of the attacked server, which now has a much larger number of requests that its network capacity cannot handle, resulting in flooding. CharGen is a protocol that creates content packets of 0-512 random characters in response to a request sent to UDP or TCP port 19. When sending a TCP request supported by the CharGen protocol, the server starts sending random characters until the connection with the server is terminated. If a UDP request is sent, the server responds with randomly selected characters every time it receives a UDP datagram [20].

In DDoS attacks, AI is used to prevent them. Artificial neural networks (ANN) are used to identify fraudulent traffic generated at the network entrance. In order to be able to use ANNs, data collection procedures containing records of traffic in the network must first be performed. The collected data goes through the process of normalizing the value parameters so that they can be used in the ANN. The next step is the creation and development of the ANN model, which includes the learning process, the evaluation process, and the testing process. An analysis of the obtained results is performed on the entire process [20]. ML can be used here when collecting data for use in ANN and during its development phase.

ML can also be used to detect DDoS attacks generated by IoT devices. IoT is a principle that depicts a network infrastructure in which any device is connected to the Internet and other devices in a network. In this network, all devices share information about themselves and the environment in which they are located [21]. The most common example of IoT networks is found in smart homes. IoT devices used in smart homes are also known as Smart Home Internet of Things (SHIoT) devices. Devices with built-in sensors are connected to an IoT network that collects data from other devices and then analyzes the collected data so that the most valuable information can be sent to applications that perform a specific task based on the analysis results [20].

The vast majority of SHIoT devices are cheap devices that do not have significant capabilities, and due to their simplicity, they are easy to compromise to perform malicious actions, including generating DDoS attacks. In order to develop an ML-based model that will detect any occurrence of anomalies in the network, there must be a dataset representing real and unreal traffic. The Logistic Model Tree (LMT) method can be used to develop such a model. LMT is a supervised machine learning classification model that combines logistic regression and Decision Tree Learning. When developing such models, SHIoT devices are divided into several classes. LMT models are then developed for each of these classes that detect anomalies, and for each class, the model is developed separately. The developed model works in two phases. The first phase is a prerequisite for detecting DDoS-generated traffic in the second phase. The first phase also includes classifying SHIoT devices based on the generated traffic [22].

Phishing attacks (identity theft) are types of fraud that most often occur via e-mail but are also common via SMS messages or phone calls. These attacks aim to steal data from a large group of users in an illegitimate way [23]. Similar to phishing are spam messages that can also be harmful to the computer and network but are mainly intended for promotions or advertisements. Phishing messages that arrive always have specific criminal purposes [24].

In Fig. 5, there is an image of an e-mail as an example of a phishing message in which the user is instructed that a specific error occurred during the payment of taxes and that in order to get his money back, he needs to create a profile via the link in the e-mail.

AI, in this example with phishing messages, can be used to divide incoming messages into spam and phishing categories based on supervised and unsupervised machine learning algorithms. The problem with distinguishing phishing messages

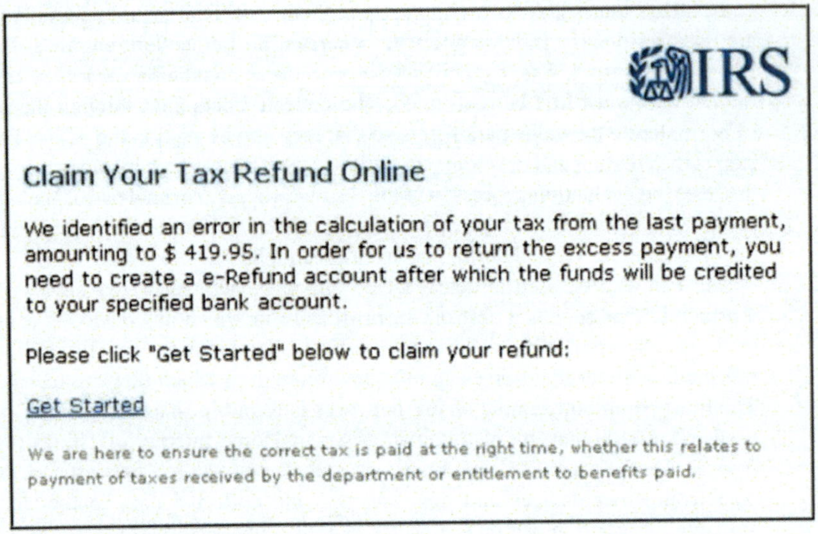

Fig. 5 Phishing message [25]

from spam messages is that they are a subtype of spam messages, so it is not easy to distinguish them. Their distinction can be made using the hierarchical clustering algorithm. The process looks like this: Incoming mail is divided into three groups. Since no marked data would be available for training, the user is asked to mark which incoming messages fall into the category of messages from the first of the three groups. When the user completes that task, the clustering algorithm has data on which to perform learning. These other messages from the other two groups are divided according to the algorithm that went through the training phase based on the user's selection. At the end of each group, the classifier from the previous group is used for the algorithm's training phase. While for the first group, the output data from the clustering algorithm are used for training [24]. In this way, the system learns to recognize a phishing message, which it can later notify the user about.

4.2 *Machine Learning in the Field of IoT Device Classification*

The type of ML used to develop the classification model is a logistic regression model with LogitBoost. LogitBoost is also used to boost the classification algorithm [26].

The first step in developing a classification model is the selection of features that will be relevant in constructing the model. The correct selection of features brings higher precision to the classification model and higher classification speed and can reduce the occurrence of excessive load. The goal is to choose the smallest number of features that will give the best results for the classification model to reduce the device class prediction time and the complexity in general [26].

The next step is to collect a set of data to develop a classification model. The goal of the model is to determine to which class that IoT device belongs. This determination can be based on the characteristics of the amount of traffic it generates in a specific period of measurement. The characteristics depend on the characteristics of the IoT device producing that traffic. Then goes the ML process in which the logistic regression algorithm is applied. The logistic regression algorithm shows the conditional probability that a specific monitored device belongs to a specific class. Then, the LogitBoost algorithm based on logistic regression is introduced, which is used for optimization. According to research [26], using this type of ML algorithm for creating a classification model proved highly accurate, with a precision of 99.79%.

4.3 Other Possibilities of Applying Machine Learning in Telecommunications and Industry 4.0

Some of the other areas of application of AI and ML in telecommunications are, for example, network traffic prediction and predictive maintenance of devices in the network. Forecasting network traffic is significant nowadays, especially when it comes to forecasting mobile network traffic because, nowadays, every day, there are more and more users of mobile networks whose management is becoming more demanding daily. Multiple ML methods can be used to predict mobile network traffic [27]. One of these methods is support vector machine (SVM). SVM is a statistical ML method whose primary goal is to map non-linear data into a higher dimensional linear space where these data will be classified by hyperplane [28]. Hyperplanes are decision boundaries that help classify data. This method can be beneficial for making network traffic congestion predictions.

Forecasting the amount of generated network traffic is done with ML, deep learning, and/or ANN algorithms. In the process of predicting the generated network traffic, ML is used in the stages of data collection, data mining, and data processing, as well as the creation of a model on which further research is then carried out. The ML algorithms that can be used in these procedures are K-means and the hierarchical clustering algorithm from the type of unsupervised machine learning. It is possible to use other ML algorithms. The choice of algorithm depends on the purpose, i.e., what one tries to achieve [29].

The smart factory environment, which aims to optimize production processes fully and is a planned outcome of Industry 4.0, is powered by AI. In addition to AI algorithms that enable computers and digital machines to accurately perform tasks related to intelligent human beings in a smart factory, ML algorithms enable predictions without explicit programming, whereby human involvement in the production process is minimal [30]. In such an environment, AI is mainly used to

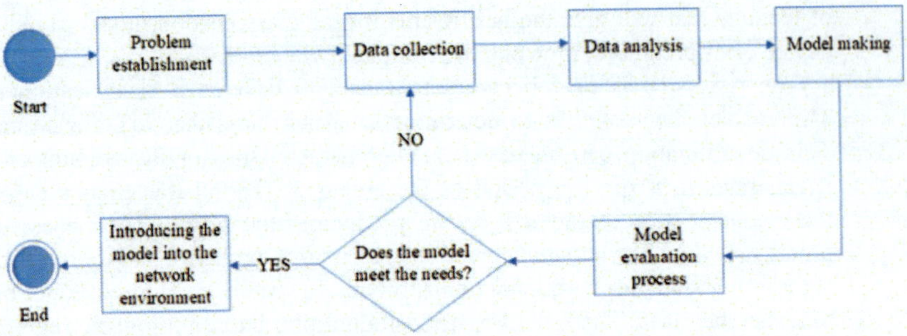

Fig. 6 UML activity diagram of machine learning in the field of telecommunications [33]

intuitively solve challenges whose formulation is entirely or not at all in codes. On the other hand, the impact of ML is visible when planning production processes, predictive maintenance, managing robotics, in help and learning systems, or process control and optimization [31].

Predictive maintenance of devices in industrial environments within the concept of Industry 4.0, Industrial IoT, and Operational Technology (OT) represents continuous monitoring of device parameters. Most often, its condition is monitored through wireless sensors on the device that send reports to the central computer responsible for further analysis of the device's condition and deciding whether the device should be replaced or repaired. Using a regression algorithm supervised by ML makes it possible to determine how many hours of practical work are left on the device. Also, with a classification algorithm of supervised machine learning, failure can be predicted in a given period [32].

Figure 6 shows the UML activity diagram of ML algorithms in telecommunications. The first step is to establish the problem because the training phase within ML is always long-term and often very expensive, and it is imperative to determine the problem of the given task accurately and precisely. The problem can be classified so that it is immediately known what type of algorithm will be used to solve it. The second step is data collection. As in the previously elaborated chapters, collecting as much data as possible to describe the network is necessary. For example, this data can be the duration of weather calls. In the third step, data analysis is done. During this step, values and features of the network are tried to be discovered in order to be able to create a model as credible as possible. The next step is model making, which includes model selection, training, and refinement, i.e., upgrading the model. In this step, care must still be taken to select the correct algorithm and method for training the model. The penultimate step in this diagram is the evaluation process in which possible errors are analyzed, and correctness cross-checked. Cross-validation is performed to see if the model works well enough. The last step is introducing the learned model into a functional network environment [33].

5 Conclusion

ML is a beneficial process nowadays because it brings greater precision to system operation, makes work easier for people who until recently had to perform these tasks independently, and now with the help of ML, they can create models and learn devices that will help them in performing the task. Although the procedure is beneficial, the complexity of the execution from start to finish should not be overlooked. From the first to the last, every step is equally vital in the ML process because if a small mistake is made at any step, the learned machine might malfunction. AI as science and technology is man's future. More and more today, people rely on the smart devices that they use and that surround them, all of which are based on AI. As a subsystem in AI, ML can be performed in three types. Of course, the best results can be expected if the device could pass through all three types of ML. Indeed, in many cases, the above is neither possible nor necessary. In practice, unsupervised machine learning is most often used because the data expected at the output is unknown in the vast majority of cases. ML has various tools and platforms in which it can be performed, and well-known programming languages such as Python, C++, Java, and similar are used to create ML code.

When talking about the possibilities of applying ML methods, it is extensive. ML can be used in almost any field. Thus, the application of ML can be found in medicine, pharmacy, the military industry, the gaming industry, and, among other things, the telecommunications industry. In the information and communication field, ML can be used for data collection and grouping, classification of IoT devices, classification of electronic mail, protection against cyber-attacks, prediction of network traffic density, and the like.

The increasing use of ML has come about due to the increased data available today. That amount of data is too much for an ordinary person and even for a team of people, and because of this, systems that enable data processing of those amounts are becoming increasingly popular. Just as everything else in the world is progressing, it is to be expected that new ML methods will appear over time, which may be simpler and more efficient than the ones that exist today.

References

1. European Commission. (2020). *White paper on artificial intelligence – A European approach to excellence and trust*. Brussels.
2. Wilson, A. *A brief introduction to supervised learning*. https://towardsdatascience.com/a-brief-introduction-to-supervised-learning-54a3e3932590. Accessed 4 June 2022.
3. Mishra, S. *Unsupervised learning and data clustering*. https://towardsdatascience.com/unsupervised-learning-and-data-clustering-eeecb78b422a. Accessed 4 June 2022.
4. Lawtomated. *Supervised learning vs unsupervised learning. Which is better?*https://lawtomated.com/supervised-vs-unsupervised-learning-which-is-better/. Accessed 4 June 2022.

5. Altexsoft. *Unsupervised learning: Algorithms and examples*. https://www.altexsoft.com/blog/unsupervised-machine-learning/. Accessed 4 June 2022.
6. Batta, M. (2020). Machine learning algorithms – A review. *International Journal of Science and Research, 9*, 381–386. https://doi.org/10.21275/ART20203995
7. Johnson, D. *Reinforcement learning tutorial*. https://www.guru99.com/reinforcement-learning-tutorial.html. Accessed 4 June 2022.
8. Brownlee, J. *Machine learning tools*. https://machinelearningmastery.com/machine-learning-tools/. Accessed 4 June 2022.
9. My server name. *11 most popular machine learning software tools 2021*. https://hr.myservername.com/11-most-popular-machine-learning-software-tools-2021. Accessed 4 June 2022.
10. Scikit-learn. *Machine learning in Python*. https://scikit-learn.org/stable/. Accessed 4 June 2022.
11. TensorFlow. *TensorFlow*. https://www.tensorflow.org/. Accessed 4 June 2022.
12. Kovalev, S., Sintsov, S., Khizhniak, A., & Turol, S. *Implementing k-means clustering with TensorFlow*. https://www.altoros.com/blog/using-k-means-clustering-in-tensorflow/. Accessed 4 June 2022.
13. PyTorch. *PyTorch*. https://pytorch.org/. Accessed 4 June 2022.
14. TutorialsPoint. *What is Weka?*https://www.tutorialspoint.com/weka/what_is_weka.htm. Accessed 1 July 2022.
15. Weka 3. *Weka 3: Machine learning software in Java*. https://www.cs.waikato.ac.nz/ml/weka/. Accessed 1 July 2022.
16. Brownlee, J. *How to run your first classifier in Weka*. https://machinelearningmastery.com/how-to-run-your-first-classifier-in-weka/. Accessed 1 July 2022.
17. RapidMiner. https://rapidminer.com/platform/. Accessed 1 July 2022.
18. Burke, J. *What is the role of machine learning in networking?*https://searchnetworking.techtarget.com/answer/What-is-the-role-of-machine-learning-in-networking. Accessed 4 July 2022.
19. Stouffer, C. *DDoS attacks: A simplified guide + DDoS attack protection tips*. https://us.norton.com/internetsecurity-emerging-threats-ddos-attacks.html. Accessed 4 July 2022.
20. Perakovic, D., Perisa, M., Cvitic, I., & Husnjak, S. (2017). Model for detection and classification of DDoS traffic based on artificial neural network. *Telfor Journal, 9*, 26–31. https://doi.org/10.5937/telfor1701026P
21. Clark, J. *What is the Internet of Things (IoT)?*https://www.ibm.com/blogs/internet-of-things/what-is-the-iot/. Accessed 4 July 2022.
22. Cvitic, I., Perakovic, D., Gupta, B. B., & Choo, K.-K. R. (2022). Boosting-based DDoS detection in internet of things systems. *IEEE Internet of Things Journal, 9*, 2109–2123. https://doi.org/10.1109/JIOT.2021.3090909
23. Phishing.org. *What is phishing?*https://www.phishing.org/what-is-phishing. Accessed 4 July 2022.
24. Gupta, B. B., Tewari, A., Cvitić, I., Peraković, D., & Chang, X. (2022). Artificial intelligence empowered emails classifier for Internet of Things based systems in industry 4.0. *Wireless Networks, 28*, 493–503. https://doi.org/10.1007/s11276-021-02619-w
25. Norton. *Phishing email examples to help you identify phishing scams*. https://us.norton.com/internetsecurity-online-scams-phishing-email-examples.html. Accessed 1 July 2022.
26. Cvitić, I., Peraković, D., Periša, M., & Gupta, B. (2021). Ensemble machine learning approach for classification of IoT devices in smart home. *International Journal of Machine Learning and Cybernetics, 12*, 3179–3202. https://doi.org/10.1007/s13042-020-01241-0
27. Jamshidi, S. (2019). *The applications of machine learning techniques in networking*.
28. Akhtar, M., & Moridpour, S. (2021). A review of traffic congestion prediction using. *Journal of Advanced Transportation, 2021*, 18. https://doi.org/10.1155/2021/8878011
29. Joshi, M., & Hadi, T. (2015). A review of network traffic analysis and prediction techniques. *ArXiv*. abs/1507.0. https://doi.org/10.48550/arXiv.1507.05722

30. Zorić, P., Musa, M., & Kuljanić, T. M. (2021). Smart factory environment: Review of security threats and risks. In *Lecture notes of the Institute for Computer Sciences, Social-Informatics and Telecommunications Engineering, LNICST* (pp. 203–214). https://doi.org/10.1007/978-3-030-78459-1_15
31. Peraković, D., Periša, M., Cvitić, I., & Zorić, P. (2021). Artificial intelligence application in different scenarios of the networked society 5.0 environment. In *Zbornik Radova Trideset Devetog Simpozijuma O Novim Tehnologijama U Poštanskom I Telekomunikacionom Saobraćaju – Postel 2021*. University of Belgrade, Faculty of Transport and Traffic Engineering. https://doi.org/10.37528/FTTE/9788673954455/POSTEL.2021.021
32. Gonfalonieri, A. *How to implement machine learning for predictive maintenance*. https://towardsdatascience.com/how-to-implement-machine-learning-for-predictive-maintenance-4633cdbe4860. Accessed 4 July 2022.
33. Wang, M., Cui, Y., Wang, X., Xiao, S., & Jiang, J. (2017). Machine learning for networking: Workflow, advances and opportunities. *IEEE Network, 32*, 92–99. https://doi.org/10.1109/MNET.2017.1700200

Guidelines for Mitigating Cybersickness During Training in VR Environment Using Head-Mounted Displays

Marko Matulin, Štefica Mrvelj, and Sergo Martirosov

1 Introduction

Ever since the dawn of computer graphics, the objective of virtual environment designers has been to immerse the users in their creations, i.e., computer-generated worlds, and improve interaction with virtual objects in them. This path, which leads to immersion, is considerably shorter if a designer can trick as many user senses as possible (audiovisual, motion, smell, touch). Since humans are predominantly visual creatures, by enveloping our visual sensory with, for instance, head-mounted display (HMD) systems used to display computer graphics, we can be led to believe that the virtual world is real. The effect is even more pronounced if the content includes spatial audio and if we interact with the virtual objects using haptic devices.

Although the gaming industry drives the development of VR technology, nowadays, this technology is used in a variety of industries and fields. Studies analyzed in this paper show how training in virtual environments can replace the hours spent in real-world training. That is very handy when training in real environments is impossible due to, e.g., high risk of injury, high cost, lack of equipment, or, more recently, lockdown measures due to a global pandemic. The technology is still developing, but it is confidently gaining momentum since it provides a cost-effective alternative for improving the trainees' professional skills.

VR can support realistic and immersive simulations and enable the transfer of taught skills in VR into real contexts and provide multi-user, embodied, and

---

M. Matulin (✉) · Š. Mrvelj
University of Zagreb, Faculty of Transport and Traffic Sciences, Zagreb, Croatia
e-mail: mmatulin@fpz.unizg.hr; smrvelj@fpz.unizg.hr

S. Martirosov
University of West Bohemia, Univerzitní 2732/8, Plzeň, Czech Republic
e-mail: martiros@rti.zcu.cz

L. Knapčíková, D. Peraković (eds.), *7th EAI International Conference on Management of Manufacturing Systems*, EAI/Springer Innovations in Communication and Computing, https://doi.org/10.1007/978-3-031-22719-6_17

interactive active learning. It is a promising tool for training and complex problem solving, which requires weighing multiple variables and situational decision-making [1, 2]. That was most recently confirmed by Narciso et al. [3], who used VR training to train professional firefighters. The study concluded that the VR training program yielded significant knowledge transfer, even more so than real-life training. The same conclusion is outlined in [4], where the authors investigated the effects of VR training on the precision of hand movements.

However, using a VR environment for gaming, training, or other purposes comes with a cost. The most obvious cost is the equipment cost, i.e., computer hardware and software used to create, distribute and present the VR content and interact with the objects inhabiting the virtual world. Yet another cost or downside of using VR is cybersickness. When working with VR applications, especially during long training sessions (e.g., close to one hour or more), the advent of cybersickness is almost inevitable. Hence, what motivates us is to reveal its origins, influential factors, and how to avoid it. The contributions of this paper are twofold. We are keen to summarize previous studies' results and then generate a concise set of guidelines that could help the industry professionals mitigate the cybersickness risk and its effects when VR and HMD systems are used for industrial training.

The paper structure is as follows. In Sect. 2, we will devote attention to briefly explaining cybersickness, while Sect. 3 provides an insight into related studies and their findings on what factors relate to the advent of cybersickness during the VR sessions. We summarize those findings into guidelines for cybersickness mitigation in Sect. 4. Section 5 concludes.

2 Cybersickness in Brief

Cybersickness is a type of motion sickness. Specifically, motion sickness appears in real life while moving (for instance, in a car, boat, rollercoaster, etc.), while cybersickness produces the same symptoms but without actual movement, since it occurs in a virtual environment. The root cause of both conditions is a mismatch between the body's motion sensors and what the eyes convey to the brain.

Cybersickness can manifest itself while looking at the moving images on a screen. For instance, while browsing through the smartphone or computer screen by scrolling it, when using multiple screens, during telemeetings when other participants control the screen, or in VR users (the latter being the focus of this paper). The affected individuals can experience nausea, sweating, dizziness, headache, and eye strain. Note that the symptoms are not to be taken lightly since aftereffects can be long-lasting, making the individuals less effective in performing other everyday tasks, such as driving. When using VR applications, the effect is more pronounced and likely to appear if the HMD system is used for content presentation, entirely enveloping a human visual system.

In the example shown in Fig. 1, a user's body senses that it is at rest and sitting down. However, the user is completely immersed in the VR scene; the HMD system

Fig. 1 VR user in a virtual race

and headphones keep the user's audiovisual system busy and overwhelmed with information. At the same time, the haptic steering wheel feeds the brain with motion signals. In turn, this can confuse the brain while interpreting different sensory inputs. The vestibular system thinks the body is resting, but the eyes see the rapid movement on HMD screens. It contradicts these expectations, leading to cybersickness.

In 2022, Gao et al. analyzed the 288,685 players' reviews of 506 VR games and concluded that cybersickness is their significant concern [5], meaning it must not be ignored when designing VR-based training programs.

3 Cybersickness Causality Factors

A wide variety of research focuses on finding the connection between virtual reality and cybersickness. However, the relationship is not as straightforward as expected, mainly because each individual reacts differently to the visual stimuli, making it difficult to predict [6]. To that end, several research endeavors are targeted at discovering the impact of human-related factors, such as age [7], gender [8], and user psychological state [9]. Specifically, Koslucher et al. showed that females are more susceptible to cybersickness than males [10]. Yet that claim was recently challenged by Stanney et al. [11]. They argued that females are more frequently experiencing the symptoms because HMD systems are usually designed for male-sized heads (i.e., poorly fitted HMD can cause cybersickness). In 2021, Melo et al. [12] also found no evidence that specific gender is more prone to developing cybersickness.

It is worth noting that VR users with a history of migraines or concussions also suffer a higher chance of getting cyber sick [10] and that sleep deprivation might strengthen the symptoms [13]. The importance of sleep was recently confirmed by Ng et al. [14], who showed that rested users had higher cybersickness tolerance. Additionally, participants with a lower pain threshold had slightly lower cybersickness tolerance and reported more symptoms.

Apart from the individual susceptibility to cybersickness, numerous technical factors play their role. Duh et al. [15] and Emoto et al. [16] both suggest that the monitor size and the viewing angle can severely influence the possibility of experiencing cybersickness. In [17], the authors conducted a study to test the impact of VR gaming session duration and found that when using the HMD system, the sickness occurred in all participants who played the game for 50 minutes. That is in line with the recent study [18], which showed that even if the short 10-minute breaks are introduced into the 50-minute VR sessions, there was no evidence that breaks can mitigate the aftereffects.

Another influential factor is latency. When rotating the head in a virtual environment with HMD on, adaptation to a newly displayed scene has to happen, which takes time. In this process, high latency and the resulting low framerates can lead to a reduced sense of presence [19], impaired task performance, reduced user response, or simulator sickness [20]. It has been determined by Adelstein et al. [21] that the critical end-to-end latency is 17 ms, but others found this threshold to be even lower [22].

Apart from latency and the resulting framerate, the field-of-view (FOV) of the virtual scene also interplays with the advent of cybersickness. Ajoy et al. [23] argued that decreasing the FOV also decreases the sickness level, yet the cost is the decrease in immersion and spatial awareness. In [24], the authors confirmed that the FOV reduction could alleviate the symptoms but did not report any drawbacks in user VR immersion when FOV is reduced. Furthermore, Naoki et al. discovered that 3D video yields higher discomfort levels than 2D video [25], while Ang et al. [26] recorded a higher level of cybersickness if the virtual environment (presented with an HMD system) consisted of regular bumps and irregular terrain (compared to traversing on a flat geometry terrain).

The VR environment characteristics analyzed in [26] are an excellent introduction to another group of cybersickness influential factors: user movement control within a VR environment. We already noted how the symptoms are experienced when a mismatch between real-life body position (movement) and the VR position of a user or its avatar exists. That means it is essential to understand how different techniques for controlling the user position in VR space interplay with cybersickness that may arise in some individuals.

In a virtual environment, a user can move around using various controllers: keyboard, mouse, joystick, gamepad, or another kind of locomotion interface (like the *bouncing seat* [27]), teleporting, body gestures (e.g., leaning the head like in [28]), or by actually moving (walking) in a real-world while a computer tracks those movements and calculates the position in the VR environment. Recently, Lin et al. [29] proposed a novel locomotion technique called head-motion assisted locomotion that can reduce the cybersickness and obtain a higher presence than teleportation. In [30], an impression of locomotion study was conducted. The authors found that VR users can be tricked into experiencing movement (specifically, walking) even if they are sitting down during the sessions. That helps avoid cybersickness issues since the users feel more stable while sitting down. Yet, it must be noted that not all industrial training programs (for specific job positions) can be completed using only sitting-down-type exercises. Some jobs require learning, for instance, moving objects around workshops or tool handling in different user postures.

4 The Mitigation Guidelines

Based on the literature review, we can now outline compact guidelines for cybersickness risk mitigation when different training programs for industry professionals are conducted in a virtual environment using HMD systems.

- Participants
  - Recent studies found no evidence that specific gender is more prone to cybersickness, i.e., both genders can participate in the training equally.
  - Training program providers should ensure that all the participants are in good health, not suffering from migraines, concussions, stability or postural issues, have good eyesight, and are well-rested before engaging in the training sessions.
- Training sessions
  - The sessions should not exceed 50 minutes at one go; ideally, they may last up to 40 minutes. 10-minute breaks can be introduced if necessary.
- Head-mounted displays
  - HMD system used for training must be well-fitted to the participants' heads, e.g., it should be lightweight and include adjustable straps making it easy to fit on different-sized heads.
  - The screens in the HMD system should have a refresh rate of 60 Hz minimum to avoid eye strain and have a high pixel density.
- Rendering device
  - A computer or other type of device used for VR environment rendering and control must be able to produce 60 frames-per-second on both HMD screens in HD resolution at all times, ensuring fluent motion no matter how fast the scene is changing.
  - If a device used for VR environment rendering is fed with the VR content from a network location (e.g., a cloud server), end-to-end latency should not exceed 17 ms.

- Virtual environment
  - When possible, the VR environment (e.g., a factory hall, workshop, warehouse, a worker's workstation, etc.) should be a flat geometry terrain without bumps or elevations.
  - High-quality, high-resolution graphics should be used for textures and VR object creation.
  - The participants' field of view may be reduced for their comfort and should include the ground level.
- Movement and object control
  - Walking in the real world should be avoided as a technique of moving around the virtual environment. Different locomotion interfaces can be used instead. Currently, promising results in preventing cybersickness are reported if body gestures are used for movement in VR.
  - If the training programs allow, the participants should be in the sitting position while immersed in VR.
  - While operating with virtual objects (for instance, different tools), it is beneficial to use haptic devices and controllers to increase the sense of immersion.

5 Conclusion

Nowadays, education via VR courses and training programs is gaining momentum. What once started as a gaming gadget gradually becomes an effective tool for professional training and skill transfer. The availability of 2D/3D/VR creation tools, such as Unreal Engine or Unity, and HMD devices only propel that growth. However, in this paper, we showed that VR also brings some drawbacks. The issue of cybersickness arises, possibly transferring the training programs into a rather unpleasant and uncomfortable experience that can lead to medical conditions like nausea, dizziness, headache, etc.

Hence, we reviewed the relevant literature to derive a set of guidelines that may help the industry professionals and the training providers mitigate the cybersickness risk and effects. The guidelines highlight the six groups of measures that can be taken to combat the condition and avoid its symptoms. It is difficult to define the rank of these groups, for instance, by their importance; however, it cannot be ignored that the locomotion interfaces used for moving around the virtual environment are currently the focus of most researchers in this field. The goal is to develop a suitable interface that: (a) would be easy to operate, (b) provides accurate haptic feedback to a user, (c) contributes to achieving a sense of immersion, (d) does not require elaborate user movement in real life, and (e) does not contribute to cybersickness level.

The outlook of our future research includes investigating viewers' cybersickness levels when watching different 360° videos using HMD devices like in [31]. We will analyze how video content types and characteristics (resolution, framerate, bitrate) affect the viewers' sickness level.

References

1. Cheng, Y., & Wang, S. H. (2011). Applying a 3D virtual learning environment to facilitate student's application ability-the case of marketing. *Computers in Human Behavior, 27*(1), 576–584. https://doi.org/10.1016/j.chb.2010.10.008
2. Bertram, J., Moskaliuk, J., & Cress, U. (2015). Virtual training: Making reality work? *Computers in Human Behavior, 43*, 284–292. https://doi.org/10.1016/j.chb.2014.10.032
3. Narciso, D., Melo, M., Rodrigues, S., Cunha, J. P., Vasconcelos-Raposo, J., & Bessa, M. E. (2022). Using heart rate variability for comparing the effectiveness of virtual vs real training environments for firefighters. *IEEE Transactions on Visualization and Computer Graphics, early access*. https://doi.org/10.1109/TVCG.2022.3156734
4. Martirosov, S., Hořejší, P., Kopeček, P., Bureš, M., & Šimon, M. (2021). The effect of training in virtual reality on the precision of hand movements. *Applied Sciences, 11*(17), 1–19. https://doi.org/10.3390/app11178064
5. Gao, Y., Chen, A., Chi, S., Zhang, G., & Hao, A. (2022). Analysis of emotional tendency and syntactic properties of VR game reviews. In *Proceedings of the conference on virtual reality and 3D user interfaces abstracts and workshops (VRW)* (pp. 648–649). IEEE. https://doi.org/10.1109/VRW55335.2022.00175
6. LaViola, J. J. (2000). A discussion of cybersickness in virtual environment. *SIGCHI Bulletin, ACM, 32*(1), 47–56. https://doi.org/10.1145/333329.333344
7. Chompoonuch, J., & Kazuhiko, H. (2011). Study on Parallax effect on simulator sickness in one-screen and three-screen immersive virtual environment. In *Proceeding of the School of Information and Telecommunication Engineering* (pp. 34–39). Tokai University.
8. Parsons, T. D., Larson, P., Kratz, K., Thiebaux, M., Bluestein, B., Buckwalter, J. G., & Rizzo, A. A. (2004). Sex differences in mental rotation and spatial rotation in a virtual environment. *Neuropsychologia, 42*, 555–562. https://doi.org/10.1016/j.neuropsychologia.2003.08.014
9. Bruck, S. R., & Watters, P. A. (2009). Cybersickness and anxiety during simulated motion: Implications for VRET. *Studies in Health Technology and Informatics, 144*, 169–173. https://doi.org/10.3233/978-1-60750-017-9-169
10. Koslucher, F., Haaland, E., & Stoffregen, T. A. (2016). Sex differences in visual performance and postural sway precede sex differences in visually induced motion sickness. *Experimental Brain Research, 234*(1), 313–322. https://doi.org/10.1007/s00221-015-4462-y
11. Stanney, K., Fidopiastis, C., & Foster, L. (2020). Virtual reality is sexist: But it does not have to be. *Frontiers in Robotics and AI*. https://doi.org/10.3389/frobt.2020.00004
12. Melo, M., Gonçalves, G., Narciso, D., & Bessa, M. (2021). Impact of different role types and gender on presence and cybersickness in immersive virtual reality setups. In *Proceedings of the international conference on graphics and interaction (ICGI)* (p. 1–8). IEEE. https://doi.org/10.1109/ICGI54032.2021.9655281
13. Sunu, W., Titis, W., Hanung, A., Muhhamad, B., & Mumtaz, N. (2015). Quantifying visual attention and visually induced motion sickness during day-night driving and sleep deprivation. In *International conference on data and software engineering* (pp. 191–194). IEEE. https://doi.org/10.1109/ICODSE.2015.7436996

14. Ng, A. K. T., Leung, C. H. Y., Chan, L. K. Y., & Lau, H. Y. (2022). K.: Human factors related to cybersickness tolerance in virtual environment. In *IEEE conference on virtual reality and 3D user interfaces abstracts and workshops (VRW)* (pp. 528–532). IEEE. https://doi.org/10.1109/VRW55335.2022.00118
15. Duh, H. B. L., Parker, D. E., & Furness, T. E. (2004). An independent visual background reduced simulator sickness in a driving simulator. *Presence: Teleoperators and Virtual Environments, 13*(5), 578–588. https://doi.org/10.1162/1054746042545283
16. Emoto, M., Sugawara, M., & Nojiri, Y. (2008). Viewing angle dependency of visually induced motion sickness in viewing wide-field images by subjective and autonomic nervous indices. *Displays, 29*(2), 90–99. https://doi.org/10.1016/j.displa.2007.09.010
17. Merhi, O., Faugloire, E., Flanagan, M., & Stoffregen, T. A. (2007). Motion sickness, console videogames, and head-mounted displays. *Human Factors, 49*(5), 920–934. https://doi.org/10.1518/001872007X230262
18. Szpak, A., Richards, A., Michalski, S. C., & Loetscher, T. (2022). Getting the most out of virtual reality: Evaluating short breaks to reduce cybersickness and cognitive aftereffects. In *IEEE conference on virtual reality and 3D user interfaces abstracts and workshops (VRW)* (pp. 533–537). IEEE. https://doi.org/10.1109/VRW55335.2022.00119
19. Slater, M., Lotto, B., Arnold, M. M., & Sanchez-Vives, M. V. (2009). How we experience immersive virtual environments: The concept of presence and its measurement. *Anuario de Psicologia, 40*(2), 193–210.
20. Buker, T. J., Vincenzi, D. A., & Deaton, J. E. (2012). The effect of apparent latency on simulator sickness while using a see-through helmet mounted display: Reducing apparent latency with predictive compensation. *Human Factors, 54*(2), 235–249. https://doi.org/10.1177/0018720811428734
21. Adelstein, B. D., Lee, T. G., & Ellis, S. R. (2003). Head tracking latency in virtual environments: Psychophysics and a model. *Human Factors and Ergonomics Society, 47*(20), 2083–2087. https://doi.org/10.1177/154193120304702001
22. Jerald. J. J. (2010). *Scene-motion and latency-perception thresholds for head mounted displays*. PhD thesis, University of North Carolina at Chapel Hill.
23. Ajoy, F., & Steven, F. (2016). Combating VR sickness through subtle dynamic field-of-view modification. In *IEEE symposium on 3D user interfaces (3DUI)* (pp. 201–210). https://doi.org/10.1109/3DUI.2016.7460053
24. Wu, F., & Rosenberg, E. S. (2022). Asymmetric lateral field-of-view restriction to mitigate cybersickness during virtual turns. In *Proceedings of the conference on virtual reality and 3D user interfaces (VR)* (pp. 103–111). IEEE. https://doi.org/10.1109/VR51125.2022.00028
25. Naoki, K., Hiroki, Y., Masahiro, I., & Yutetsu, M. (2015). Effects of visual induced motion sickness of stereoscopic 3D interactive video. In *4th global conference on consumer electronics (GCCE)* (pp. 664–665). IEEE. https://doi.org/10.1109/GCCE.2015.7398678
26. Ang, S., & Quarles, J. (2022). You're in for a bumpy ride! Uneven terrain increases cybersickness while navigating with head mounted displays. In *Proceedings of the conference on virtual reality and 3D user interfaces (VR)* (pp. 428–435). IEEE. https://doi.org/10.1109/VR51125.2022.00062
27. Onuki, Y., & Kumazawa, I. (2022). Bouncing seat: An immersive virtual locomotion interface with LSTM based body gesture estimation. In *Proceedings of the conference on virtual reality and 3D user interfaces abstracts and workshops (VRW)* (pp. 834–835). IEEE. https://doi.org/10.1109/VRW55335.2022.00268
28. Hashemian, A. M., Lotfaliei, M., Adhikari, A., Kruijff, E., & Riecke, B. E. (2022). HeadJoystick: Improving flying in VR using a novel leaning-based interface. *IEEE Transactions on Visualization and Computer Graphics, 28*(4), 1792–1809. https://doi.org/10.1109/TVCG.2020.3025084
29. Lin, Z., Gu, X., Li, S., Hu, Z., & Wang, G. (2022). Intentional head-motion assisted locomotion for reducing cybersickness. *IEEE Transactions on Visualization and Computer Graphics, early access*. https://doi.org/10.1109/TVCG.2022.3160232

30. Saint-Aubert, J., Cogne, M., Bonan, I., Launey, Y., & Lecuyer, A. (2022). Influence of user posture and virtual exercise on impression of locomotion during VR observation. *IEEE Transactions on Visualization and Computer Graphics, early access*. https://doi.org/10.1109/TVCG.2022.3161130
31. Mrvelj, Š., Matulin, M., & Martirosov, S. (2020). Subjective evaluation of user quality of experience for omnidirectional video streaming. *PROMET Traffic & Transportation, 32*(3), 421–433. https://doi.org/10.7307/ptt.v32i3.3444

Research on the Influence of the Digital Transformation on Operations Management Practices: Challenges and Future Directions

Anna Saniuk and Dagmar Caganova

1 Introduction

The digital transformation (DT) gives a chance to huge increase productivity and innovation and also brings changes in the global economy in consumption, investment, growth, employment, and trade [1, 2]. Without doubt, it leads to a leap in efficiency and increases the effectiveness of value chains as well as helps to reduce energy and resource consumption [3, 4]. DT has driven the new line of research, because it generates many technical as well as management problems in industrial enterprises, especially nowadays, when progressive digitalization, automation, dynamic development of mobile wireless networks, and new advanced technologies change the way of business activities [5–8]. Although most studies have been devoted to the technological aspects of DT, rather little attention has been paid to upcoming changes in operations management. Besides, there is rather insufficient information about the way that top managers are currently applying the digital transformation in their own everyday practices and what their expected outcomes are. A number of research questions have been set: How the digital transformation is changing an operations management? To what extent should management change in the real industrial company in the digital transformation environment? Which barriers have practitioners encountered?

The overall goal of this work is to investigate how the digital transformation influences on operations management practices. For this aim, the qualitative data was collected through an expert interview study to identify the key expected

A. Saniuk (✉)
University of Zielona Góra, Zielona Góra, Poland
e-mail: a.saniuk@wez.uz.zgora.pl

D. Caganova
Comenius University in Bratislava, Faculty of Management, Bratislava, Slovakia

L. Knapčíková, D. Peraković (eds.), *7th EAI International Conference on Management of Manufacturing Systems*, EAI/Springer Innovations in Communication and Computing, https://doi.org/10.1007/978-3-031-22719-6_18

outcomes of the digital transformation in operations management. The article concentrates on tangible good production and presents the research results based on the individual in-depth interviews (IDI) with intentionally selected experts and top managers from Polish medium industrial enterprises to derive practical insights. This research proposal is a part of a wider debate and should help to investigate the possibility of applying existing and creating new management methods for a successful digital transformation.

The remainder of this paper includes five sections. Section 2 contains the literature review. Section 3 identifies design methods and the study area selection. Section 4 shows the qualitative research results, formulates the study findings, and discusses the results of previous research. Section 5 defines the conclusions and indicates the directions of future research.

2 Literature Review

2.1 Main Digital Manufacturing Technologies

The digital transformation observed on the global market is based on digital manufacturing technologies (DMTs) represented by especially: cloud computing, big data, Internet of things, blockchain technology, artificial intelligence (AI), 5G, 3D printing, cybersecurity, autonomous robots, simulations, digital twins, additive manufacturing, machine learning, augmented and virtual reality, etc. [9–11]. The evidence seems to be strong that the implementation of DMTs to improve processes is a need of the day. Nevertheless, the study shows that enterprises mainly concentrate in Just-in-Time manufacturing and Technology Management [12].

On the one hand, the implementation of Internet of things, cloud computing, and big data technologies gives very big opportunities for industrial companies through a significant increase in flexibility and efficiency, and better control over operations and supply chain [13]. Mass personalization, mass customization, and simultaneous reduction of manufacturing costs are possible. However, on the other hand, it is also a huge challenge especially for SMEs, which represent about 90% of businesses worldwide [7, 14]. Several researchers underline that the main strengths of small and medium enterprises are especially short time of decision making, high flexibility, and good employee relations and cooperation, while the greatest weaknesses are infrastructural and technical constraints and small financial resources [15–22].

The significant object of DT is the cyber-physical system (CPS) which means a set of machines, robots, IT systems, workpieces, vehicles, and devices equipped with technologies, e.g., smart sensors, biosensors, RFIDs, telematics, microprocessors or complete embedded systems which are flexible and self-controlling. They can collect and process data about themselves and their surroundings and also asses data. However, the significant advantage is that CPS can communicate with each other and different systems. All production areas should be fully integrated and also controlled using advanced IT technologies [23–25]. In addition, the real-time connection of physical and digital systems is necessary [26]. Interconnection among systems and different technologies such as cloud computing, sensors, radio frequency identifier, and cyber-physical systems is also necessary [12].

The Industry 4.0 and Industry 5.0 concepts support DT and assume the application of many different technologies which transform the industrial production. Some of them are already used in manufacturing, especially in large enterprises, but they can create together integrated, advanced, optimized, and automated production flow with much higher productivity. CPS is connected along the value chain with other systems by Internet-based protocols and allows data to be processed and analyzed which is used to quickly adapt to changes, configure themselves, and even to predict failures in advance. It leads to many changes in relationships between manufacturers, suppliers, customers, and also between human and machine [27–29]. Otherwise, it can significantly improve business and production processes of the enterprise [30]. In this transformation, connected CPS allows a process flexibility and efficiency to be increased, a quality of goods to be improved, and at the same time the unit cost to be reduced. The manufacturing productivity and competitiveness of companies can be higher [31, 32].

Nowadays, the digital transformation is a recent trend and a part of industrial strategy allows connecting with partners, customized products, enhancing the work environment, achieving autonomous processes, and synchronizing flows [27]. Many studies presented in the literature have stated that digital manufacturing technologies enable to monitor manufacturing processes, and improve their flexibility and capabilities. Moreover, they pointed out that whereas cloud computing technology and RFID are rather often used in practice, more expensive and revolutionary technologies such as big data, collaborative robot, CPS, blockchain, and machine-to-machine are still disused. A lack of experts, resources, research and development departments, and complex IT tools are highlighted by many authors of research [33].

2.2 *Changes in Operations Management in the Digital Transformation Environment*

In a dynamically changing environment and intense global competition, the ability of manufacturing organizations to identify and seize opportunities in the market becomes the basis for a competitive advantage [34]. The market existence of today's enterprises depends on the quick response and adaptation of the market offer to constantly changing customer expectations [33]. The Covid-19 pandemic confirmed these trends. An effective constant searching for new chances and occasions in the market to achieve a competitive advantage, high organizational flexibility and productivity, a quick response, and short decision-making processes are priorities in the management of modern companies.

In addition to technological changes, the digital technologies radically change the structure and all management of the company at an internal level as well as in relationships [31, 35–37]. It means strategic and operational changes and also changes in relations between manufacturers and customers [38]. It is clear that the digital transformation has a positive influence and creates new chances for enterprises [39]. Several studies described in the literature have shown that scientists and practitioners most often conduct research on technological changes in companies, rather too less attention has been paid to changes in production and management [40].

Operations management (OM) can be defined as a discipline which should be used in problem solving in the management of the manufacturing industry [41]. OM is described as planning, designing, and controlling of the manufacturing process and "redesigning business operations in the production of products or services" [42]. OM is also defined as "managing work to produce valuable results" [43] or as "management of the transformation process that converts labor, capital, materials, information, and other inputs into products and services for customers" [44].

DT involves high costs of implementing new technologies, the need to strictly control costs, high pressure on reducing manufacturing costs, and shortening the period of return on investment. A lack of resources and knowledge to implement modern and advanced technologies are also barriers for companies [45]. As research shows, mostly industrial enterprises are limited to introducing only Cloud Computing and Internet of Things. However, they do not introduce a real transformation of the business models [33, 46]. The previous study pointed out that there is a need to build new business models in the Industry 4.0 era [47].

The digital manufacturing fulfills individual customer needs through a greater diversity of products manufactured in small quantities and even one-off items in real time. It does not need to reconfigure an assembly line and setup times. The industrial enterprises have to change from existing forms of business activity to the new platform for much greater efficiency, flexibility, and speed. They should create a smart factory oriented toward high value added production and high profitability of production plants. This requires a huge transformation of organizations and their processes. The process should be managed in a new way and a different approach to management is required [26]. A digital thinking, collecting, and analyzing data are important elements of a success [24, 25, 48].

It must be claimed that the organizational changes play important role in the digital transformation. As a result of DT, a "different" enterprise is created, which requires a different approach to management and a well-thought-out formulated ad hoc strategy that included different aspects of the company's operations, and also changed relationships between clients, suppliers, producers, and employees [36]. The major drivers for DT investments should be: (1) a reduction in operating costs; (2) an improvement of manufacturing speed; (3) an improvement of production quality; and (4) an improvement of process quality [40, 49, 50]. As the research indicates, enterprises which introduce the digital technologies must change their strategy and overcome many barriers, whereas the most important of them include: technical infrastructure, automation of devices which can communicate between them, employee skills, and collaboration with other companies in the aim to build together network of companies – smart factories of the future [51, 52].

The previous studies about the DT clearly highlight a need to change the approach to business models [17, 53–55], which mean a description of how the company operates [56]. There are three approaches to business models: financial, operational, and strategic. The financial approach refers to making decisions regarding the financial aspects of the company and indicates how the company generates profits. The operational approach mentions internal processes that are involved in creating value, while the strategic approach is related to aspects of company positioning, its development opportunities, and cooperation with other entities.

The business model should lead to creating a balanced competitive advantage of the company on a given market, based on both the financial and operational as well as strategic aspects [57]. Innovations in business models provide financial and business benefits by reducing costs and improving flexibility in adapting to changing market needs and by avoiding or removing risks from competitors [58]. Customers expect customized products tailored to their individual needs according to the mass customization concept. As a result of the dynamic development of the Internet, the producer has direct contact with the client. New possibilities of personalizing the purchased products and obtaining information directly from customers about the use of products and their future needs arise. The digital technology tools require a high level of expertise and a large investment as well, but allow more flexibility to be achieved due to a decentralization of information and decision-making process [28, 59]. To reduce costs new business models should be created such as, for example, product-as-a-service, in which investment costs are limited by replacing them with operating costs such as leasing, subscription, lease, etc. Data analysis or management of a machine park also can be treated as a service. The customized products should be characterized by the well-coordinated combination of products and services [60–62]. Companies have to be highly flexible. Customers expect customized or even personalized products adapted to human necessities. The results of the present studies also highlight that the digital servitization seems to be a current trend in the market and means "an integration of products and services" [63–66].

3 Materials and Methods

3.1 Research Purpose

The paper is on the context of the digital transformation of industrial enterprises and its impact on operations management. The aim of this study is to investigate how the digital transformation of industrial enterprise influences operations management practices. For achieving this aim, the research questions have been set, which are listed below:

- What is changing as an effect of digital transformation in industrial enterprises?
- Which of these changes may trigger the need for changes in operations management to improve an economic efficiency?
- How to manage industrial enterprises to eliminate problems and threats within DT environment?
- What changes in operations management need to be made to more effectively implement DT?

Effectiveness is understood here as a result of actions taken, thanks to which obtained effects exceed the expenditure incurred [67]. The conducted research sought to improve an economic efficiency which means increasing in work efficiency, productivity of fixed assets, investment, and production efficiency [68]. The conducted research is based on medium industrial enterprises which belong to the SMEs sector. Small and medium enterprises (SMEs) mean the enterprises which "employ less than 250 employees and have an annual turnover not exceeding 50 million EUR and/or an annual balance sheet total not exceeding 43 million EUR" according to the EU recommendation 2003/361 [69].

3.2 Research Methodology

To answer the questions, the following research methods were conducted:

Literature Review An attempt is made to structure the problem area through analysis of existing needs and challenges for operations management of industrial enterprises in DT environment. The previous studies are focused mainly on the new technologies of DT. However, they do not introduce a completely new approach to operations management and a real transformation of the business model. It is clear therefore that there is a need to conduct research about upcoming changes in operations management. Besides, the literature reviews recognize many symptoms suggesting that a new approach to operations management in DT age in practice is needed.

Individual in-depth interviews (IDI) with selected experts and operations managers In light of the aforementioned gaps, the qualitative research was conducted to identify how the digital transformation influences on operations management practices. The individual in-depth interview analysis was used. This technique enables to conduct intensive interviews with specially selected experts and top managers from Polish medium industrial companies to identify upcoming needed changes in the management process from their point of view and based on their practical insights.

Firstly, the aim and the scope of the qualitative research method were explained. The IDI research was designed to be semi-structured and use open-ended questions to discuss from the digitalization and advanced technologies to their influence on industrial enterprises. After that, the respondents were asked to identify their implications on operations management and next to describe the exact areas of changes. Then, follow-up and spontaneous questions were used to stimulate discussion and deepen the accuracy of the detailed changes described in operations management and their cause-effect relationship with the digital transformation.

The interviews were conducted in Poland from September 2021 to April 2022. Such a long period of research was associated with the restrictions related to the pandemic Covid-19. IDI were conducted in intentionally selected 35 industrial medium enterprises in Poland. The companies were purposefully selected for interviews. They had to meet all of the following criteria: (1) belonged to the SME sector; (2) were based in Poland; (3) have operated in the manufacturing sector for at least 6 years. The individual in-depth interviews were made with specially selected experts and operations managers of these enterprises. For each company one employee was deliberately selected (35 people in total). In this research, the interviewers (operations managers) were the critical determinant of the data. They had to demonstrate interdisciplinarity, knowledge, and practice of both management issues and advanced technologies. The results of the study were described in three stages:

- Step 1 – identification of key factors that trigger the need to change the existing solutions related to operations management.
- Step 2 - identification, what areas of OM should be changed.
- Step 3 – identification, what detailed changes in OM should be made.

The interviews were not recorded for three reasons. Firstly, many respondents refused to be interviewed for the recording. Secondly, it was found that the conversation without recording was freer and less stressful for the respondents, which favored a more accurate identification of changes in OM. Besides, interviewers feel looser and say exactly what they think [70]. This is confirmed by studies that the recording device which is used to record the conversation has an effect on the data [71]. Thirdly, research indicates that the quality of data obtained by recording the interview is similar to the quality of data obtained from taking notes immediately after the interview and can be also the best approach [72]. The most important information, critical points expressed by the operations managers were written down during the conversation and detailed notes were prepared immediately afterward.

4 Research Results and Discussion

As a result of step 1, key factors, which have influence on operations management in the future, were reported by the operations managers based on their extensive practical experience (Fig. 1). The most frequently indicated factor was real-time communication (RTC) mode. Thirty-three respondents highlighted an importance of this factor and its influence on operations management. Especially, they expect to increase a role of both the real-time analyzing data streams, real-time control as well as decision-making policies. The previous research confirmed these opinions,

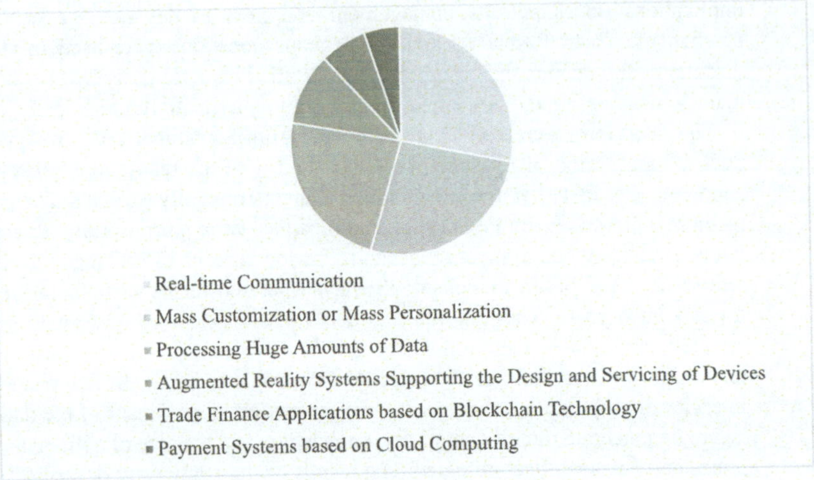

Fig. 1 Key factors that have an influence on operations management in the digital transformation environment. (Source: own study based on IDI results)

that the popularization of the Internet of things caused a need of processing and analyzing data streams in real-time, which is a new challenge and is growing rapidly [73–76]. The significant role and generation of data in all productive processes and using them as a basis for decision making were pointed out in previous research [77–78].

As many as 31 respondents identified mass customization or mass personalization as the important ones. Mass customization (MC) means a kind of production strategy in which enterprises connect customer and enterprise needs and offer a mass-customized product at a low price [79–81]. The hybrid make-to-stock and make-to-order manufacturing occurs when producers offer a great diversity of products with satisfactory lead time [82]. Mass personalization (MP) can be defined as a kind of production paradigm in which producers offer customers many mass customized products and services and also they give the customer the opportunity to order these products from anywhere in the world and at any time and they deliver them in the way that customers choose [83].

Twenty-seven experts recognized processing huge amounts of data as a key factor. Besides, 12 respondents revealed that augmented reality system supporting the design and servicing of devices is a factor which can have an influence on operations management due to shorten time of design and servicing and new opportunities for planning. Eight respondents claimed that trade finance applications based on blockchain technology are the key factor and six respondents indicated payment systems based on cloud computing.

Table 1 The result of IDI in SMEs – identification of changes in the production planning

Area of changes	Expected effects – IDI results
Production planning	Continuous and automatic supplementation and updating of data on manufactured variants of products
	Appreciable decrease in time of each product variant realization (up to 24/36 hours)
	Generation of automatic material orders taking into account the inventory in real-time
	Guaranteed order fulfillment within a few hours
	Appreciable decrease in time of product design
	Finite number of product variants
	Planning ready-made procedures for the implementation of each possible product variant
	Efficient enterprise resource planning
	Real-time production planning
	Real-time production scheduling
	Real-time enterprise resource planning
	Real-time production changes
	Leagile strategy – more often used
	Agile strategy – more often used
	Priority process automation according to the bottleneck
	Continuous procedure planning oriented on reduction of bottlenecks

Source: own study based on IDI results

Then, IDI were analyzed by developing a list of significant areas of operations management. As a result of step 2, the operations managers identified planning, staffing, controlling, and organizing as the main areas of changes in operations management which are the effect of the digital transformation.

In step 3, the respondents have revealed the detailed changes which are expected in near future within highlighted areas of operations management. The interviews clearly indicated a need of changes in production planning, which is a result of mass customization or mass personalization (Table 1). Mass personalization production means that a variety of retail channels and delivery methods should be applied – traditional retail, outlets, etc., and also ordering by computer, mobile phones, etc. [83]. The experts stated a necessity to introduce next-day or same-day delivery. They highlighted an importance of real-time communication mode.

Mass customization should enable the customers to choose based on their individual needs regarding the product appearance, features, and needs related to their use, and to analyze the product life cycle to constantly improve the product. Therefore, respondents pointed out that production planning process must enable, on the one hand, the production of highly personalized products, on the other hand, it must drastically reduce the time of design, planning, and production preparation to allow production immediately after customer order. Thus, the selection of variants of the manufactured product must be defined and prepared in such a way as to avoid the need for time-consuming product design by preparing multiple versions of the designs in advance, and even the production of product components facilitating the rapid production of customer-tailored product variants.

What is interesting, 9 experts proposed to use the strategy leagile and 7 experts proposed the agile strategy to help preparing much faster products for quick assembly according to customer needs. Based on the IDI, it was found that the customized products must be considered simultaneously with a strong pressure to reduce production costs (29 respondents) and shorten the deadline for preparing the order for execution (22 respondents).

The experts concluded that the product variants possible to choose should be characterized by:

- Low production costs (20 experts)
- Short time for preparing the order for execution (17 experts)
- Production capacity of the enterprise (13 experts)
- Production system bottleneck (11 experts)
- The possibility of material delivery according to the Just-in-Time principle (5 experts).

Similarly, the scientists based on previous research indicated a need of changes in scheduling and production planning and control [84, 85]. DT enables to make decisions of production planning and control by faster and efficient information process helping to reduce the cost, increase profits and enhance operational efficiency [86, 87]. Besides, the researchers also inferred that it is necessary to change a delivery mode. Nowadays, customers expect next-day or even same-day delivery. In the future will be expected a same-hour delivery. Retailers must deliver mass personalized products using many different channels, orders, and delivery methods [83, 88]. The meaning of the bottleneck was also indicated in other manuscripts [89]. The personalized products increase enterprise financial results, however only to some degree, beyond which they decrease what was indicated by M. Ben-Jebara and S. B. Modi [90]. However, H.K.S. Lam, L. Ding, and Z. Dong stated that foreign competition has less influence on product differentiation than on cost leadership strategy [91]. Based on the literature analysis, scientists pointed out that new technological innovation and digitization enable among others agile manufacturing [92].

Secondly, the interviews have clearly shown a need of changes within controlling (27 respondents), what is presented in Table 2. The introduction of the Industry 4.0 concept is resulted in digitization, automation and robotization, which are very cost-intensive. Therefore, constant cost control and a pressure to reduce costs are needed. The operations managers claimed that one of the effects of Industry 4.0 is a significant increase in indirect costs (24 respondents). They stated that a new costing method is required (17 respondents). It is expected faster cost measurement of different, personalized products in real-time (12 respondents). Many of respondents (8) have also reported that the high costs of new technologies are related to the need to control and monitor the return on investment, which is expected by owners/shareholders and management of the company.

Table 2 The result of IDI in SMEs – identification of needs in controlling

Area of changes	Expected effects and needs – IDI results
Cost measurement	Increase in indirect costs More accurate settlement of indirect costs More accurate cost control New cost model Replacing average cost on precisely calculated cost Product unit cost determined on the basis of the cost of participation in individual processes Step reduction in the time of cost calculation of product variants Continuous real-time updating of costs of individual processes Continuous cost estimation of individual processes
Strategy	Fast implementation of the strategy Frequent introduction of changes Use of procedures that allow for immediate changes in OM after the introduction of new strategic goals Strong pressure to introduce new technologies Constant elements of the strategy linked with the Industry 4.0
Pricing	The possibility of pricing a product variant based on continuously updated data on costs Real-time price calculation Price calculation based on real-time changes in material costs

Source: own study based on IDI results

Using the IDI interview technique, respondents were asked about the possibility of cost measurement with the currently used costing method after the implementation of Industry 4.0. Majority of experts (31) agreed that using the currently costing method in their companies it is not possible to determine the unit product cost in real-time. The average product unit cost based on information about the unit cost of similar products (very approximate) can be calculated. Also, 29 respondents claimed that it is not possible to estimate the unit product cost using the current costing method in their enterprise when products differ only in finish, color, lead time, packing time, transport costs, assembly time, etc. Nine experts proposed to use the activity-based costing (ABC) method, which allows indirect costs for individual product variant and cost of individual process to be detailed calculated. The scientists based on previous research have expressed a similar view. A need of more careful cost control, measurement, and a new cost model is reported in many articles [84, 89, 93].

Besides, 22 practitioners expect a need of changes in strategy and 18 operations managers in pricing. The respondents claimed that strategy should be oriented on quick changes (24 experts) and new technologies (17 experts). Moreover, strategic goals require fast implementation into enterprise practice in the light of very changeable environment (14 experts). Similarly, the literature highlighted the crucial role of a strategy in enterprise [91].

Table 3 The result of IDI in SMEs – identification of changes in the organizing area

Area of changes	Expected effects – IDI results
Organizational Structure (OS)	Flatter OS OS focused around processes OS focused on quick changes OS focused on innovation Short decision-making process Simple procedures Less bureaucracy

Source: own study of IDI results

The interviews' participants clearly recognized that fundamental changes during the implementation of new technologies of the Industry 4.0 are needed in the organizing area (Table 3), especially within an organizational structure (18 respondents). The organizational structure should be focused on processes (17 respondents), innovation (13 respondents), and quick changes (9 respondents). It was also noticed that the decision-making process should be shortened (12 respondents), especially through simple procedures (9 respondents) and less bureaucracy (5 respondents).

Moreover, the respondents underlined that it is needed a flatter structure – a horizontal structure (16 respondents). Therefore, using the IDI interview technique, the most important benefits of introducing a horizontal structure to the company were determined, based on the opinions of experts from these companies who already had experience in the functioning of such a structure (9 experts in total):

- Easier information flow (8 experts)
- Increasing the flexibility of the company through better possibilities of coordinating all activities focused on processes (7 experts)
- Increased responsibility of teams (5 experts)
- Information flow in both directions (4 experts)
- Increased employees' sense of responsibility for the results of the introduced changes (4 experts)

The literature analysis confirmed that a result of digitization in the Industry 4.0 era are changes within organizational structure [94, 95]. Many scientists reported that a flat hierarchy within organizational structure as well as decentralized decision-making is expected [59, 96–98].

Key changes are required in the staffing area that operations managers revealed (see Table 4). The interdisciplinary knowledge of employees and a necessity of interdisciplinary teamwork are very strongly highlighted in IDI research (all 35 respondents). The experts recognized that a greater responsibility and independence of employees are needed (28 respondents), especially due to more new and unexpected problems resulting from the implementation of the advanced digital technologies (22 respondents). Furthermore, a wide range of employee tasks has been stated (19 respondents), which explains, among others, the interdisciplinary nature of employees.

Table 4 The result of IDI in SMEs – identification of changes in the staffing area

Area of changes	Expected effects – IDI results
Human resources (HR)	Significantly greater interdisciplinarity of employees Motivation system oriented toward introducing new technologies Procedures which allow easy replacement of employees Precise division of tasks Greater responsibility and independence of employees Preparation for a wide range of employee tasks Greater pressure to perform tasks Clear procedures Clear division of responsibilities More teamwork Interdisciplinary team Greater responsibility of teams Employees' responsibility for the execution of orders Great independence of employees Increased responsibility for employees' decisions

Source: own study of IDI results

Likewise, prior studies demonstrated a need of interdisciplinary knowledge of employees, interdisciplinary team creation, and quick adaptation to frequent changes [59, 92]. The necessity of multi-disciplinary approach, new skills, and capability was underlined by G. Culot, G. Nassimbeni, G. Orzes, and M. Sartor [99, 100].

5 Conclusions

The aim of this research was to define the upcoming effects of the digital transformation on operations management in industrial enterprises. The study is based on a qualitative approach and reported results of the individual in-depth interviews with specially selected experts and operations managers based on their extensive experience. The interviews included three stages and identified significant factors that are expected to trigger a need for changes, main areas of changes in OM, and description of detailed changes which should be made.

The following significant conclusions and recommendations can be drawn from the present research about trends and upcoming changes in operations management based on the results and outcomes of qualitative research:

1. The study has found that generally the changes in OM caused by the digital transformation are expected in planning, controlling, organizing, and staffing areas.
2. Taken together, the operations managers have identified that the main reasons for upcoming changes are a result of rapidly evolving technologies and increasing

opportunities arising from real-time communications and a possibility to process huge amounts of data. They have stated the development of mass customization and mass personalization as the next main factors. Besides, they revealed that the augmented reality systems supporting the design and servicing of devices, trade finance applications based on blockchain technology and payment systems based on cloud computing have a great influence on changes in operations management.

3. The results of these interviews have shown that an industrial company in the digital transformation environment should react rapidly to changes. Therefore, the organizational structure should be flatter and also fast implementation of strategy and much greater interdisciplinarity of employees are needed.
4. The respondents of this study pointed out that mass customization and mass personalization lead to a huge diversity of products in smaller quantities. The customers can choose variants of products and expect an immediate delivery. Therefore, producers have to organize production in such a way that the assembly and delivery should be realized no more than about 36 hours. This causes the need for changes in planning, developing the new cost models, and changes in pricing based on real-time mode.
5. The extensive experience with the digital transformation of operations managers has shown that all time-consuming processes must be prepared in advance, and even many of them to a large extent, which causes the need for changes in the field of planning and controlling.
6. One of the more significant findings to emerge from this study is that the employees must be much more interdisciplinary, work in interdisciplinary teams, and take more responsibility for their actions.

Overall, the better part of the findings of this investigation complements earlier research. These obtained study results are relevant and should be of wide interest for both academics and practice experts. The revealed reasons and expected outcomes of the digital transformation in operations management enable a conceptual framework for future studies in the scope of new business models supported the digital transformation.

The findings from the presented research make twofold contribution to the current literature. First, without doubt, several changes in operations management were identified which can be very useful and allow a new approach to OM in the digital transformation environment to be built. Second, a formulation of expected changes helps to assess and modify existing or create new management methods in conditions of the developing digital transformation. By shedding light on these results, scientists and practitioners are able to device better methods and solutions for improving the management of manufacturing SMEs. Besides, the findings of this study can help to build a solid base for improving the strategy and a reference point to inspire scientists to conduct further research in this direction.

References

1. Pedota, M., & Piscitello, L. (2021). A new perspective on technology-driven creativity enhancement in the Fourth Industrial Revolution. *Creativity and Innovation Management*. https://doi.org/10.1111/caim.12468
2. Schwab, K. (2016). *The Fourth Industrial Revolution*. World Economic Forum. ISBN 9781944835002.
3. Kagermann, H. (2015). Change through digitization – Value creation in the age of Industry 4.0. In H. Albach, H. Meffert, A. Pinkwart, & R. Reichwald (Eds.), *Management of Permanent Change* (pp. 23–45). Springer Gabler.
4. Marre, M., Beihofer, D., Haggenmueller, W. A., & Grupp, P. (2015). Forming for resource-efficient Industry 4.0. In *Proceedings of the 5th international conference on accuracy in forming technology (ICAFT), Chemnitz, Germany, 10-11 November 2015* (pp. 655–673).
5. Demartini, M., & Taticchi, P. (2021). Performance measurement and management. A literature review focused on the role played by management theories with a deep dive into the industry 4.0 environment. *International Journal of Productivity and Performance Management*.https://doi.org/10.1108/IJPPM-02-2021-0063
6. Bai, R., Chandra, V., Richardson, R., & Liu, P. P. (2020). Next generation mobile wireless networks: 5G cellular infrastructure. *Journal of Technology Management and Applied Engineering, 36*(3), 1–19.
7. Serumaga-Zake, J. M., & van der Poll, J. A. (2021). Addressing the impact of Fourth Industrial Revolution on South African manufacturing small and medium enterprises (SMEs). *Sustainability, 13*(21), 11703.
8. Vial, G. (2019). Understanding digital transformation: A review and a research agenda. *The Journal of Strategic Information Systems, 28*(2), 118–144.
9. Rocha, C., Quandt, C., Deschamps, F., & Philbin, S. (2022). R&D collaboration strategies for industry 4.0 implementation: A case study in Brazil. *Journal of Engineering and Technology Management, 63*, 101675. https://doi.org/10.1016/j.jengtecman.2022.101675
10. Peng, B. (2021). Digital leadership: State governance in the era of digital technology. *Culture of Science. SAGE*, 1–16. https://journals.sagepub.com/doi/pdf/10.1177/2096608321989835. Accessed 21 July 2022
11. Gillani, F., Chatha Kamran, A., Sadiq Jajja, M. S., & Farooq, S. (2020). Implementation of digital manufacturing technologies: Antecedents and consequences. *International Journal of Production Economics, 229*, 107748.
12. Fettermann, D. C., Cavalcante, C. G. S., de Almeida, T. D., & Tortorella, G. L. (2018). How does Industry 4.0 contribute to operations management? *Journal of Industrial and Production Engineering, 35*(4), 255–268.
13. Gastaldi, L., Lessanibahri, S., Tedaldi, G., & Miragliotta, G. (2022). Companies' adoption of smart technologies to achieve structural ambidexterity: An analysis with SEM. *Technological Forecasting and Social Change, 174*, 121187.
14. Small and Medium Enterprises (SMEs) Finance. *Improving SMEs' access to finance and finding innovative solutions to unlock sources of capital*.https://www.worldbank.org/en/topic/smefinance. Accessed 8 July 2022
15. Yang, H. X., & Ren, W. R. (2021). Research on the influence mechanism and configuration path of network relationship characteristics on SMEs' innovation-the mediating effect of supply chain dynamic capability and the moderating effect of geographical proximity. *Sustainability, 13*(17), 9919.
16. Garcia-Muina, F. E., Medina-Salgado, M. S., Ferrari, A. M., & Cucchi, M. (2020). Sustainability transition in Industry 4.0 and smart manufacturing with the triple-layered business model canvas. *Sustainability, 12*(6), 2364.
17. Svatosova, V. (2021). Proposal and simulation of a business process model of strategic management in E-commerce. *Ekonomicky Casopis, 69*(7), 726–749.

18. Feifei, J., Ying, Z., Harish, G., Jinpei, L., & Jia, C. (2021). Evaluation of small and medium-sized enterprises' sustainable development with hesitant fuzzy linguistic group decision-making method. *Applied Intelligence*.https://doi.org/10.1007/s10489-021-02372-9. Accessed 14 July 2022.
19. Dangayach, G. S., & Deshmukh, S. G. (2001). Manufacturing strategy: Literature review and some issues. *International Journal of Operations & Production Management, 21*(7), 884–932. https://doi.org/10.1108/01443570110393414
20. Grando, A., & Belvedere, V. (2006). District's manufacturing performances: A comparison among large, small-to-medium-sized and district enterprises. *International Journal of Production Economics, 104*(1), 85–99.
21. Stevenson, M., Huang, Y., Hendry, L. C., & Soepenberg, E. (2011). The theory and practice of workload control: A research agenda and implementation strategy. *International Journal of Production Economics, 131*(2), 689–700.
22. Land, M. J., & Gaalman, G. J. C. (2009). Production planning and control in SMEs: Time for change. *Production Planning & Control, 20*(7), 548–558. https://doi.org/10.1080/09537280903034230
23. Andronie, M., Lazaroiu, G., Stefanescu, R., Uta, C., & Dijmarescu, I. (2021). Sustainable, smart, and sensing technologies for cyber-physical manufacturing systems: A systematic literature review. *Sustainability, 13*, 5495. https://doi.org/10.3390/su13105495
24. Bochmann, L., Gehrke, L., Böckenkamp, A., Weichert, F., Albersmann, R., Prasse, C., Mertens, C., Motta, M., & Wegener, K. (2015). Towards decentralized production: A novel method to identify flexibility potentials in production sequences based on flexibility graphs. *International Journal of Automation Technology, 9*(3), 270–282.
25. Thoben, K. D., Wiesner, S., & Wuest, T. (2017). "Industrie 4.0" and smart manufacturing – A review of research issues and application examples. *International Journal of Automation Technology, 11*(1), 1–16.
26. Olsen, T. L., & Tomlin, B. (2020). Industry 4.0: Opportunities and challenges for operations management. *Manufacturing & Service Operations Management, 22*(1), 113–122.
27. Hwang, B. G., Ngo, J., & Teo, J. Z. K. (2022). Challenges and strategies for the adoption of smart technologies in the construction industry: The case of Singapore. *Journal of Management in Engineering, 38*(1), 05021014.
28. Rüßmann, M., Lorenz, M., Gerbert, P., Waldner, M., Justus, J., Engel, P., & Harnisch, M. (2015). *Industry 4.0: The future of productivity and growth in manufacturing industries* (p. 9). Boston Consulting Group.
29. Raguseo, E., Vitari, C., & Pigni, F. (2020). Profiting from big data analytics: The moderating roles of industry concentration and firm size. *International Journal of Production Economics, 229*, 107758.
30. Blunck, E., & Werthmann, H. (2017). Industry 4.0 – An opportunity to realize sustainable manufacturing and its potential for a circular economy. *Dubrovnik International Economic Meeting, 3*(1), 644–666.
31. Gastaldi, L., Lessanibahri, S., Tedaldi, G., & Miragliotta, G. (2022). Companies' adoption of smart technologies to achieve structural ambidexterity: An analysis with SEM. *Technological Forecasting and Social Change, 174*(1), 121187. https://doi.org/10.1016/j.techfore.2021.121187
32. Gerbert, P., Lorenz, M., Rüßmann, M., Waldner, M., Justus, J., Engel, P., & Harnisch, M. (2015). *Industry 4.0: the future of productivity and growth in manufacturing industries*. Boston Consulting Group. https://www.bcg.com/publications/2015/engineered_products_project_business_industry_4_future_productivity_growth_manufacturing_industries.aspx. Accessed 18 July 2021
33. Moeuf, A., Pellerin, R., Lamouri, S., Tamayo-Giraldo, S., & Barbaray, R. (2018). The industrial management of SMEs in the era of Industry 4.0. *International Journal of Production Research, 56*(3), 1118–1136. https://doi.org/10.1080/00207543.2017.137264
34. Si, S., Zahra, S. A., Wu, X., & Jeng, D. J. F. (2020). Disruptive innovation and entrepreneurship in emerging economics. *Journal of Engineering and Technology Management, 58*, 101601.

35. Felsberger, A., & Reiner, G. (2020). Sustainable Industry 4.0 in production and operations management: A systematic literature review. *Sustainability, 12*, 7982.
36. Piccarozzi, M., Aquilani, B., & Gatti, C. (2018). Industry 4.0 in management studies: A systematic literature review. *Sustainability, 10*(3821), 1–24.
37. Pan, M., Sikorski, J., Kastner, C. A., Akroyd, J., Mosbach, S., Lau, R., & Kraft, M. (2015). Applying Industry 4.0 to the Jurong Island Eco-industrial Park. *Energy Procedia, 75*, 1536–1541.
38. Piątek, Z. (2019). *Od Industry 4.0 do Smart Factory: technologie i zmiany organizacyjne*, http://przemysl-40.pl/index.php/2018/02/01/od-industry-4-0-do-smart-factory-czesc-2/. Accessed 21 July 2022
39. Gaspar, M., & Juliao, J. (2020). *Impacts of Industry 4.0 on operations management: Challenges for operations strategy*. IEEE7th international conference on industrial engineering and applications. April 16-21, 2020, Bangkok, Proceedings Paper pp. 82–87.
40. Felsberger, A., Qaiser, F. H., Choudhary, A., & Reiner, G. (2022). The impact of Industry 4.0 on the reconciliation of dynamic capabilities: Evidence from the European manufacturing industries. *Production Planning & Control, 33*(2-3), 277–300. https://doi.org/10.1080/09537287.2020.1810765
41. Peinado, J., Graeml, A. R., & Vianna, F. (2018). Operations management body of knowledge and its relevance to manufacturing and service organizations. *Operations Management Body of Knowledge, 25*(4), 373–389.
42. Helmold, M., & Terry, B. (2021). Operations Management 4.0. In *Operations and Supply Management 4.0. Future of Business and Finance*. Springer. https://doi.org/10.1007/978-3-030-68696-3_2. Accessed 22 July 2022
43. Browning, T. R. (2020). Operations management writ large. *Journal of Operations Management, 66*(5), 494–500.
44. Hill, A. V. (2011). *The encyclopedia of operations management: A field manual and glossary of operations management terms and concepts*. FT Press.
45. Hansen, E. B., & Bogh, S. (2021). Artificial intelligence and internet of things in small and medium-sized enterprises: A survey. *Journal of Manufacturing Systems, 58*, 362–372.
46. Mittal, S., Khan, M. A., Romero, D., & Wuest, T. (2018). A critical review of smart manufacturing & Industry 4.0 maturity models: Implications for small and medium-sized enterprises (SMEs). *Journal of Manufacturing Systems, 49*, 194–214.
47. Blunck, E., & Werthmann, H. (2017). *Industry 4.0 – An opportunity to realize sustainable manufacturing and its potential for a circular economy*. Dubrovnik International Economic Meeting, Sveučilište u Dubrovniku, pp. 644–666.
48. Kumar, K., Zindani, D., & Davim, J. P. (2020). *Design thinking to digital thinking*. Springer. ISBN: 978-3-030-31359-3.
49. Tsai, W. H., & Lai, S. Y. (2018). Green production planning and control model with ABC under industry 4.0 for the paper industry. *Sustainability, 10*, 2932.
50. Micieta, B., Binasova, V., Lieskovsky, R., Krajcovic, M., & Dulina, L. (2019). Product segmentation and sustainability in customized assembly with respect to the basic elements of industry 4.0. *Sustainability, 11*, 6057.
51. Saniuk, A., & Saniuk, S. (2017). *Management of the metallurgical enterprise according to the Industry 4.0 concept*. 26th international conference on metallurgy and materials - METAL 2017, TANGER, Czech Republic, Brno, pp. 2034–2039, ISBN: 978-80-87294-79-6.
52. Saniuk, S., Samolejova, A., Saniuk, A., & Lenort, R. (2015). Benefits and barriers of participation in production networks in a metallurgical cluster – Research results. *Meta, 54*(3), 567–570.
53. Gerlitz, L. (2016). Design management as a domain of smart and sustainable enterprise: Business modelling for innovations and smart growth in Industry 4.0. *The International Journal Entrepreneurship and Sustainability Issues, 3*(3), 244–268.
54. Turber, S., vom Brocke, J., Gassman, O., & Fleisch, E. (2014). Designing business models in the era of Internet of Things. In M. C. Tremblay, D. VandeeMeer, M. Rothenberger, A. Gupta, & V. Yoon (Eds.), *Advancing the impact of design science: Moving from theory to practice* (pp. 17–31). Springer International Publishing.

55. Burmeister, C., Luttgens, D., & Piller, F. T. (2015). *Business model innovation for Industry 4.0: Why the "Industrial Internet" mandates a new perspective on innovation*. RWTH Aachen University, Technology and Innovations Management.
56. Arend, R. (2013). The business model: Present and future – Beyond a skeuomorph. *Strategic Organisation, 11*(4), 390–402.
57. Morris, M., Schindehutte, M., & Allen, J. (2005). The entrepreneur's business model: Toward a unified perspective. *Journal of Business Research, 58*, 726–735.
58. IBM Global Technology Services. (2006). Business model innovation – the new route to competitive advantage, http://www-935.ibm.com/services/uk/cio/flexible/enflex_wp_business_model_innovation.pdf. Accessed 17 July 2022
59. Veile, J. W., Kiel, D., Muller, J. M., & Voigt, K. (2020). Lessons learned from Industry 4.0 implementation in the German manufacturing industry. *Journal of Manufacturing Technology Management, 31*(5), 977–997.
60. Shamn, K., Cang, S., Yu, H., & Li, Y. (2016). *Management approaches for Industry 4.0 a human resource management perspective*. IEEE congress on evolutionary computation, 24-29 July 2016, Vancouver, Canada, Proceedings Paper.
61. Li, Y., Dai, J., & Cui, L. (2020). The impact of digital technologies on economic and environmental performance in the context of industry 4.0: A moderated mediation model. *International Journal of Production Economics, 229*, 1–13.
62. Qi, Y. N., Mao, Z. G., Zhang, M., & Guo, H. F. (2020). Manufacturing practices and servitization: The role of mass customization and product innovation capabilities. *International Journal of Production Economics., 228*, 107747.
63. Ferreira, R., Scur, G., & Nunes, B. (2021). Preparing for smart product-service system (PSS) implementation: An investigation into the Daimler group. *Production Planning & Control*. https://doi.org/10.1080/09537287.2020.1821402. Accessed 21 July 2022.
64. Kamal, M. M., Sivarajah, U., Bigdeli, A. Z., et al. (2020). Servitization implementation in the manufacturing organisations: Classification of strategies, definitions, benefits and challenges. *International Journal of Information Management, 55*, 102206.
65. De la Calle, A., Freije, A., & Oyarbide, A. (2021). Digital product-service innovation and sustainability: A multiple-case study in the capital goods industry. *Sustainability, 13*(11), 6342.
66. Shah, S. A. A., Jajja, M. S. S., Chatha, K. A., & Farooq, S. (2020). Servitization and supply chain integration: An empirical analysis. *International Journal of Production Economics, 229*, 107765.
67. Begg, D., Fischer, S., & Dornbusch, R. (1995). *Ekonomia I*. PWE.
68. Encyklopedia popularna PWN. (2017). Wydawnictwo Naukowe PWN, Warszawa
69. European Commission, Entrepreneurship and Small and Medium-sized Enterprises (SMEs). https://ec.europa.eu/growth/smes/business-friendly-environment/sme-definition_pl. Accessed 16 July 2022
70. Oakley, A. (2016). Interviewing women again: Power, time and the gift. *Sociology, 50*(1), 195–213.
71. Nordstrom, S. N. (2015). Not so innocent anymore: Making recording devices matter in qualitative interviews. *Qualitative Inquiry, 21*(4), 388–401.
72. Rutakumwa, R., Mugisha, J. O., Bernays, S., Kabunga, E., Tumwekwase, G., Mbonye, M., & Seeley, J. (2020). Conducting in-depth interviews with and without voice recorders: A comparative analysis. *Qualitative Research, 20*(5), 565–581.
73. Dubuc, T., Stahl, F., & Roesch, B. (2020). Mapping the big data landscape: Technologies, platforms and paradigms for real-time analytics of data streams. *IEEE Access, 9*, 15351–15374. https://doi.org/10.1109/ACCESS.2020.3046132. Accessed 10 July 2022.
74. Shim, J. P., & Nisar, K. (2021). *Real-time streaming technology and analytics for insights*. Digital innovation and entrepreneurship, 27th annual Americas conference on information systems (AMCIS2021), August 9-13, 2021, Proceedings Paper.
75. Fernandes, E., Salgado, A. C., & Bernardino, J. (2020). *Big data streaming platforms to support real-time analytics*. 15th international conference on software technologies (ICSOFT), Proceedings, pp. 426-433.

76. Liu, Y., Zhu, Q., & Seuring, S. (2020). New technologies in operations and supply chains: Implications for sustainability. *International Journal of Production Economics, 229*, 107889. https://doi.org/10.1016/j.ijpe.2020.107889
77. Chen, F., Deng, P., Wan, J., Zhang, D., Vasilakos, A. V., & Rong, X. (2015). Data Mining for the Internet of Things: Literature review and challenges. *International Journal of Distributed Sensor Networks, 11*(8) https://journals.sagepub.com/doi/full/10.1155/2015/431047
78. Saucedo-Martinez, J. A., Perez-Lara, M., Marmolejo-Saucedo, J. A., Salais-Fierro, T. E., & Vasant, P. (2018). Industry 4.0 framework for management and operations: A review. *Journal of Ambient Intelligence Humanized Computing, 9*, 789–801.
79. Hwang, J., Kim, S., & Lee, Y. (2021). Mass customization in food services. *International Journal of Hospitality Management, 93*, 102750.
80. Franke, N., & Piller, F. (2004). Value creation by toolkits for user innovation and design: The case of the watch market. *Journal of Production Innovation Management, 21*(6), 401–415.
81. Jost, P. J., & Susser, T. (2020). Company-customer interaction in mass customization. *International Journal of Production Economics, 220*, 107454.
82. Pereira, D. F., Oliveira, J. F., & Carravilla, M. A. (2022). Merging make-to-stock/make-to-order decisions into sales and operations planning: A multi-objective approach. *Omega-International Journal of Management Sciences, 107*, 102561.
83. Wang, Y., Ma, H., Yang, J., & Wang, K. (2017). Industry 4,0: A way from mass customization to mass personalization production. *Advanced Manufacturing, 5*, 311–320.
84. Alami, D., & Maraghy, W. (2021). A cost benefit analysis for industry 4.0 in a job shop environment using a mixed integer linear programming model. *Journal of Manufacturing Systems, 59*, 81–97.
85. Zhang, L. L., Lee, C. K. M., & Akhtar, P. (2020). Towards customization: Evaluation of integrated sales, product and production configuration. *International Journal of Production Economics, 229*, 1–13.
86. Li, A. Q., Kumar, M., Claes, B., & Found, P. (2020). The state-of-the-art of the theory on Product-Service Systems. *International Journal of Production Economics, 222*, 107491.
87. Nguyen, T., Li, Z., Spiegler, V., Ieromonachou, P., & Lin, Y. (2018). Big data analytics in supply chain management: A state-of-the-art literature review. *Computers & Operations Research, 98*, 254–264.
88. Alim, M., & Beullens, P. (2020). Joint inventory and distribution strategy for online sales with a flexible delivery option. *International Journal of Production Economics, 222*, 107487.
89. Cotrino, A., Sebastian, M. A., & Gonzalez-Gaya, C. (2020). Industry 4.0 roadmap: Implementation for small and medium-sized enterprises. *Applied Sciences-Basel, 10*(23), 8566.
90. Ben-Jebara, M., & Modi, S. B. (2021). Product personalization and firm performance: An empirical analysis of the pharmaceutical industry. *Journal of Operations Management, 67*(1), 82–104.
91. Lam, H. K. S., Ding, L., & Dong, Z. (2022). The impact of foreign competition on domestic firms' product quality: Evidence from a quasi-natural experiment in the United States. *Journal of Operations Management.*, 1–22. https://doi.org/10.1002/joom.1169
92. Letmathe, P., & Rossler, M. (2022). Should firms use digital work instructions? Individual learning in an agile manufacturing settings. *Journal of Operations Management, 68*(1), 94–109.
93. Zhichao, L., Yong, Y., & Ya, C. (2021). Management and control of economic cost and analysis of investment effect in small and medium-sized enterprises under the background of Industry 4.0. *International Journal of Technology Management, 84*(3-4), 211–228.
94. Fettig, K., Gacic, T., Koskal, A., Kuhn, A., & Stuber, F. (2018). *Impact of Industry 4.0 on organizational structures*. IEEE international conference on engineering technology and innovation. 17-20 June 2018, Stuttgart, Germany, Proceedings Paper.
95. Gasparetto, W., Egger, G., Giusti, A., Rauch, E., Riedl, M., & Matt, D. T. (2018). Intelligent workpiece carrier for distributed data collection and control in manufacturing environments. *Procedia Manufacturing, 24*, 190–195.

96. Margues, M., Agostinho, C., Zacharewicz, G., & Jardim-Goncalves, R. (2017). Decentralized decision support for intelligent manufacturing in Industry 4.0. *Journal of Ambient Intelligence and Smart Environments, 9*(3), 299–313.
97. Margues, M., Agostinho, C., Zacharewicz, G., Poler, R., & Jardim-Gocalves, R. (2018). Responsive production in manufacturing: A modular architecture. In V. Sgurev, V. Jotsov, & J. Kacprzyk (Eds.), *Practical issues of intelligent innovations, studies in systems decision and control* (Vol. 140, pp. 231–254).
98. Ozkan-Ozen, Y. D., & Kazancoglu, Y. (2021). Analysing workforce development challenges in the Industry 4.0. *International Journal of Manpower*. https://doi.org/10.1108/IJM-03-2021-0167
99. Culot, G., Nassimbeni, G., Orzes, G., & Sartor, M. (2020). Behind the definition of the Industry 4.0: Analysis and open questions. *International Journal of Production Economics, 226*, 107617.
100. Saniuk, S., Caganova, D., & Saniuk, A. (2021). Knowledge and skills of industrial employees and managerial staff for the Industry 4.0 implementation. *Mobile Networks and Application*. https://doi.org/10.1007/s11036-021-01788-4

EAI MMS 2022 – 7th EAI International Conference on Management of Manufacturing Systems, October 5–7, 2022, Krynica-Zdrój, Poland

1 Steering Committee

1.1 Chair

Imrich Chlamtac
Bruno Kessler Professor, University of Trento, Italy

1.2 Members

Prof. Dragan Perakovic, Ph.D.
University of Zagreb, Faculty of Transport and Traffic Science, Croatia
A/Prof. Lucia Knapcikova
Technical University of Kosice, Faculty of Manufacturing Technologies, Slovakia
Assoc. Prof. Dr. Michal Balog
Technical University of Kosice, Faculty of Manufacturing Technologies, Slovakia

2 Organizing Committee

2.1 General Chairs

Assoc. Prof. Lucia Knapcikova
Technical University of Kosice, Faculty of Manufacturing Technologies, Slovakia
Prof. Dragan Perakovic
University of Zagreb, Faculty of Transport and Traffic Science, Croatia

L. Knapčíková, D. Perakovič (eds.), *7th EAI International Conference on Management of Manufacturing Systems*, EAI/Springer Innovations in Communication and Computing, https://doi.org/10.1007/978-3-031-22719-6

2.2 Technical Program Committee Chairs

Assoc. Prof. Lucia Knapcikova
Technical University of Kosice, Faculty of Manufacturing Technologies, Slovakia
Prof. Dragan Perakovic
University of Zagreb, Faculty of Transport and Traffic Science, Croatia
Prof. Franz Haas
TU Graz, Austria
Assist. Prof. Annamaria Behunova
Technical University of Košice, Faculty of Manufacturing Technologies, Slovakia
Dr. Ivan Cvitic
University of Zagreb, Faculty of Transport and Traffic Science, Croatia
Assoc. Prof. Dr. Michal Balog
Technical University of Kosice, Faculty of Manufacturing Technologies, Slovakia
Assoc. Prof. Marko Periša
Faculty of Transport and Traffic Sciences, University of Zagreb, Croatia
Shakib Zohrehvandi, Ph.D.
Kharazmi University, Tehran, Iran
Prof. Darina Ondrušová, Ph.D.
Alexander Dubček University of Trenčín, Faculty of Industrial Technologies in Púchov, Slovakia
Prof. Marianna Pajtášová, Ph.D.
Alexander Dubček University of Trenčín, Faculty of Industrial Technologies in Púchov, Slovakia
Assoc. Prof. Dorota Czarnecka-Komorowska, Ph.D.
Poznan University of Technology, Poland
Eng. MSc. Justyna Trojanowska, Ph.D.
Poznan University of Technology, Poland
Eng. MSc. Waldemar Matysiak, Ph.D.
Poznan University of Technology, Poland
Eng. MSc. Jakub Hajkowski, Ph.D.
Poznan University of Technology, Poland
Markus Brillinger, Dipl.-Ing. Dr.techn.
Pro2Future, GmbH., Graz, Austria

2.3 Web Chair

Dr. Jozef Husar
Technical University of Kosice, Faculty of Manufacturing Technologies, Slovakia

2.4 Publicity and Social Media Chairs

Assoc. Prof. Lucia Knapcikova
Technical University of Kosice, Faculty of Manufacturing Technologies, Slovakia
Prof. Dragan Perakovic
University of Zagreb, Faculty of Transport and Traffic Science, Croatia

2.5 Workshops Chair

Dr. Jozef Husar
Technical University of Kosice, Faculty of Manufacturing Technologies, Slovakia

2.6 Sponsorship & Exhibits Chairs

Assoc. Prof. Lucia Knapcikova
Technical University of Kosice, Faculty of Manufacturing Technologies, Slovakia
Prof. Dragan Perakovic
University of Zagreb, Faculty of Transport and Traffic Science, Croatia

2.7 Publications Chairs

Assoc. Prof. Lucia Knapcikova
Technical University of Kosice, Faculty of Manufacturing Technologies, Slovakia
Prof. Dragan Perakovic
University of Zagreb, Faculty of Transport and Traffic Science, Croatia

2.8 Panels Chair

Prof. Jan Pitel
Technical University of Kosice, Faculty of Manufacturing Technologies, Slovakia

2.9 Local Chair

Assoc. Prof. Lucia Knapcikova

Technical University of Kosice, Faculty of Manufacturing Technologies, Slovakia

3 Technical Program Committee Chairs

Lucia Knapcikova
Technical University of Košice, Faculty of Manufacturing Technologies, Slovak Republic
Dragan Perakovic
Faculty of Transport and Traffic Sciences, University of Zagreb, Croatia
Annamária Behúnová
Technical University of Košice, Faculty of Manufacturing Technologies, Slovak Republic
Franz Haas
TU Graz, Austria
Ivan Cvitic
Faculty of Transport and Traffic Sciences, University of Zagreb, Croatia
Michal Balog
Technical University of Košice, Faculty of Manufacturing Technologies, Slovak Republic
Marko Periša
Faculty of Transport and Traffic Sciences, University of Zagreb, Croatia

3.1 Technical Program Committee Members

Alessandro Ruggiero
University of Salerno, Italy
Annamária Behúnová
Technical University of Košice, Faculty of Manufacturing Technologies, Slovak Republic
Dagmar Caganova
Faculty of Materials Science and Technology, Slovak Technical University, Slovak Republic
Goran Marković
University of Belgrade, Serbia
Ivan Grgurević
Faculty of Transport and Traffic Sciences, University of Zagreb, Croatia
Janusz Grabara
Czestochowa Univeristy of Technology, The Management Faculty, Poland
Jerzy Winczek
Czestochowa University of Technology, Poland
Jozef Husár

Technical University of Košice, Slovak Republic
Marko Periša
University of Zagreb, Faculty of Traffic and Transport Sciences, Croatia
Oleg Zabolotnyi
Lutsk National Technical University, Ukraine
Vitalii Ivanov
Sumy State University, Ukraine

Index

L. Knapčíková, D. Peraković (eds.), *7th EAI International Conference on Management of Manufacturing Systems*, EAI/Springer Innovations in Communication and Computing, https://doi.org/10.1007/978-3-031-22719-6